The Good

Katie Wood is a name many travellers are familiar with. As a well established travel writer and journalist, she has tackled everything from backpacking in Europe, in her bestseller *Europe by Train*, to the deluxe world of country house hotels and continent-hopping in the *Round the World Air Guide*. With eighteen successful guidebooks to her name, and a string of freelance commissions, from the *Guardian*, *Independent* and *Scotsman* to the *Observer* and *Traveller* magazine, she is well qualified to comment on all aspects of the travel industry. She also regularly broadcasts on travel for the BBC and ITV

The perfect marriage of minds, from this series' point of view, has resulted from her marriage to Syd House. Syd is a graduate of ecological sciences and has considerable experience of practical management in conservation. Working as a Forest District Manager for the Forestry Commission, he is responsible for the management and conservation of a large part of Scotland, from Fife to the Angus glens. He has travelled the world in his own right, before combining his skills with those of his wife and looking at the impact of tourism on the environment. Together they bring you the definitive work on 'green travel', examined from both the tourist and the environmental points of view.

Syd and Katie are both graduates of Edinburgh University and share a Scottish background. Katie was born in Edinburgh. After graduation she worked in journalism, and following an eighteen-month spell backpacking in Europe, her first travel guide was published in 1983. A Fellowship from the Royal Geographical Society, several travel consultancies and fifty-seven countries later she is still writing and travelling full-time. Syd, originally from Gourock, travelled the world before returning to Scotland, where he became immersed in conservation matters and resumed managerial work within the Forestry Commission. He too is a Fellow of the Royal Geographical Society.

The couple married in 1984. They have two young sons, Andrew and Euan, live in Perth, Scotland, and lead a lifestyle punctuated with foreign travel, midnight writing, and an active involvement in conservation matters. They have recently completed consultancy work for, among others, the English Tourist Board, working on a policy document on green tourism.

By the same authors

Europe By Train
The Round the World Air Guide
European City Breaks
The Best of British Country House Hotels
Holiday Ireland
Holiday Scotland
The Cheap Sleep Guide to Europe
The 100 Greatest Holidays in the World*
The 1992 Business Travel Guide*
The Good Tourist in the UK*
The Good Tourist in France*

Also available from Mandarin Paperbacks

The Good Tourist

A Worldwide Guide
for the
Green Traveller

**KATIE WOOD
& SYD HOUSE**

Mandarin

A Mandarin Paperback
THE GOOD TOURIST

First published in Great Britain 1991
by Mandarin Paperbacks
Michelin House, 81 Fulham Road, London SW3 6RB
This edition published 1992

Mandarin is an imprint of the Octopus Publishing Group,
a division of Reed International Books Limited

Copyright © 1991, 1992 by Katie Wood and Syd House
The authors have asserted their moral rights.

A CIP catalogue record for this title
is available from the British Library
ISBN 0 7493 1136 3

Typeset by Falcon Typographic Art Ltd,
Edinburgh & London
Printed and bound in Great Britain
by Cox & Wyman Ltd, Reading, Berks

This book is sold subject to the condition
that it shall not, by way of trade or otherwise,
be lent, resold, hired out, or otherwise circulated
without the publisher's prior consent in any form
of binding or cover other than that in which
it is published and without a similar condition
including this condition being imposed
on the subsequent purchaser.

*To our sons, Andrew and Euan,
and all travellers of the next generation.*

Research/Editorial Assistants

Elizabeth Archer
Joyce Greenwood
Donald Greig
Vicky Lewis

Contents

	Acknowledgements	viii
	Preface	ix
1	The State of Play: Choices and Dilemmas	1
2	The Tourist Industry	12
3	The Negative Impact of Tourism	27
4	Tourism, the Provider?	50
5	The Way Forward	61
6	The Good Tourist	98
7	Conservation Holidays	118
8	Voluntary Work Holidays	132
9	Wildlife/Ecological Holidays	145
10	Alternative Holidays	171
11	Activity/Adventure Holidays	196
12	Conventional Holidays	223
	Chapter Notes	235
	Further Reading	238
	Appendix – Contact Organisations	240

Acknowledgements

We wish to acknowledge the help of Alison Stancliffe, founder member of Tourism Concern; James McCarthy of the Nature Conservancy Council; Peter Hughes of 'Wish You Were Here' for his effective help; Dick Sisman of Green Flag International; Sue Ockwell of A.I.T.O.; and lastly the various Tourist Boards and tour operators who answered our lengthy questionnaires.

Thanks too to Jane Carr for getting this on the shelves on time, and on the home front a big thanks to Katie's mum for helping out with our two very little lads whilst we burned the midnight oil to meet what looked like an impossible deadline.

Preface

The world is getting smaller, and though it may sound paradoxical, people are travelling further. Who would have thought, even one generation ago, that tourism and travel would be the largest industry of the 1990s? More revenue is generated through tourism and its related services than from any other source of wealth in the world economy. That is more than from oil, more than from the automobile industry, and more than the chemical industry.

Whilst most of us think of tourism as a 'good thing' in that it gives people enjoyment and pleasure, encourages cultural exchange and understanding between different races, and brings in money and accordingly raises living standards, the truth is that there is a 'hard' side to tourism – a side that destroys the natural environment and social structure of the host country. Feelings of envy are generated in those who contrast their own Third World lifestyles with the opulent tourists who come to their land to lie on the beach and consume a round of drinks at a price that would keep a family there alive for a week. How much of the wealth that tourism generates in this situation reaches *them* in their everyday life? How much cultural (and, indeed, physical) prostitution can these societies be expected to take before seeing the tourist industry as a latter day colonial power, over which they have no control and which adversely affects their lives in far-reaching ways?

The physical impact of mass tourism on the environment is another controversial area. The Mediterranean Costas, with their attendant concrete jungles, sewage and destruction of habitat, are losing popularity by the season, as the disadvantages of this glamorous industry become obvious. The spectacular Alps, winter preserve of the few for many years, are now under threat due to the millions of skis and the associated infrastructure that mass tourism has brought in its wake.

The Sphinx is crumbling to dust under the pollution of the late twentieth century, not aided by the coachloads of tourists disembarking every ten minutes; Venice is sinking under the weight of tourists; the Parthenon is roped off and pollution is affecting its stonework badly; the Lake District has had to be closed in summer due to over-crowding. All these are familiar headlines, so what is the solution? Ban all mass tourism? Impose strict government controls? Limit who goes on holiday, and when? Tourists themselves have not been slow to recognise the pitfalls of their industry. Not that they feel tourism is their industry at all, or has anything to do with them. No one accepts their role as a 'tourist'; 'traveller' perhaps, but tourists are other people, aren't they? We all joke about going to a Costa, meeting the neighbours, eating fish and chips and drinking English beer, and as this concept becomes more pronounced, and the Costas lose their appeal, a new breed of 'traveller' is emerging. Going independent, travelling further into the interior, choosing somewhere 'unspoilt', and demanding more: more ethnic experiences, more genuine culture, more understanding of the people they meet. And they don't want to harm the environment they travel to. Unfortunately, as is proved by places such as Thailand and Goa, this trend can lead to equally dangerous development, but it is this dilemma that we are addressing in this book. In our efforts to be 'good tourists', are we inevitably bound to fail? Is the nature of the beast such that we are bound to destroy what we set out to see? Can we ever properly appreciate a country, when we only have two weeks, and are staying in a purpose-built place, run for tourists, and when contact with the locals is limited to seeing them as waitresses and cab drivers?

We are not so pessimistic. We believe it *is* possible to be a Good Tourist, and this is not a contradiction in terms. We explore the issues surrounding this subject and end with practical suggestions for 'positive holidays'. Tourism is a fact of life. It's here to stay, and what it now needs is proper controls, proper planning and travellers who do not leave their consciences at home, but travel with an enlightened approach and an open mind.

The objectives in writing this book are to provide a balanced analysis of the good and bad aspects of tourism; to offer tips and advice to prospective tourists; to promote a more responsible form of tourism – the 'good tourist' concept – and finally to provide information on the range of holidays available that conform to the basics of 'good tourism'.

We are aiming the book at the general tourist travelling outside the UK for his or her holiday, although many of the principles are, we hope, applicable to the domestic as well as to the international tourist. Undoubtedly there is already a strong groundswell of opinion wanting to change certain aspects of tourist development. We hope to interest not only those people, but also the greater mass of tourists who may not yet be aware of the issues confronting them.

Each one of us has to take responsibility for tourism; all our holidays count. Your impact is as great as the next person's. We are *all* tourists, whether we like to think of ourselves as 'travellers' because we choose to be backpackers in the Himalayas, or whether we prefer to sun ourselves on a Spanish beach.

In the writing and researching of this work, we have had, on the whole, an encouraging response from the industry and from the various governments and bodies we have contacted. Our hope is that there will be an annual new edition of the book, bringing you up to date with the general travel scene. A series of guides to individual countries has now been launched; starting with *The Good Tourist in the UK* and *The Good Tourist in France*. Aimed at intelligent, caring travellers who want to get the most out of their trips, while not spoiling what they set out to see, we hope this will be the start of a new type of guidebook for a new breed of tourist.

Katie Wood and Syd House

Chapter 1

The State of Play: Choices and Dilemmas

Global awareness of environmental issues is growing at an explosive rate. The proliferation of organisations – voluntary, action and special interest groups – concerned with such matters shows clearly the increase in public concern. This interest is not confined to local or even national issues, but, increasingly, is on a global scale. Governments and businesses have had to respond, realising that environmental quality is an increasingly high priority in most people's lives.

Tourism as a major economic and environmental activity is not immune to this trend and concern. The world's biggest single economic activity has attracted its own fair share of criticism, as unease has spread at the scale of development and lack of restrictions and controls. Tourists, of all people, are highly aware of the quality of the environment they spend good money travelling to. Increasingly they are making their voices heard – this book is part of that process.

The Good Tourist series – this book and its individual country guides – is all about helping you to see that there *are* things you can do to ensure your holiday is a positive experience, both for yourself and for the country, people and environment of your destination.

In the last forty years tourism and travelling by Britons has changed enormously in both volume and nature. Just one generation ago foreign travel for pleasure was a luxury for the privileged few who had the necessary leisure and wealth. Now, in the 1990s, tourism is the world's largest single industry, generating more revenue than any other business in the world economy. This dramatic pattern of growth is set to continue well into the future.

Almost everyone in the developed world can afford to travel and few people have no established once-a-year holiday. All in all, it's very much a growth industry and therefore it's not surprising that it has been perceived as being *the thing* to go into to get rich quick. Governments and individuals alike have seen tourism as their great opportunity.

As an industry generating international tourism receipts of $230 billion in 1990[1] tourism is obviously a powerful force. Individually we gain enjoyment, relaxation and cultural exchange from our holidays. Living in a highly industrialised, materialistic and competitive society, we feel the need to make the most of our holidays as a chance to recuperate from the stresses of daily life. We travel in order to 'get away from it all'. The chance to escape from routine and to relax is vital to our mental and physical well-being. Or at least, so we perceive it. Sadly, all too often we fail to consider this aspect when planning our holidays and plunge ourselves into situations which, far from reducing stress, actually aggravate the condition. How many times have you wished for another holiday to recover from the after-effects of the first?

Tourism is without doubt a valuable economic commodity able to create wealth, give employment and endow status at both local and national levels. It stimulates development not only in the destination country, through the provision of employment, facilities and accommodation, but also in the home-base of the tourist through the travel and organisation of holidays. Developing or Third World countries in particular can benefit from increasing interest shown by Western visitors searching for different and more unusual holidays. Ideally, such countries will develop their tourism in such a way that it directly benefits their own economy. Often it is those features which cost the locals nothing – scenery, weather, landscape – which attract the visitors and bring in a much needed source of revenue. Tourists and travellers can also benefit from such development since it provides them with the chance to cross geographical and cultural borders that only twenty years ago were closed to them. As ever more distant corners of the world become accessible so we have the chance to learn more about the customs and culture of the people that share this world with us. Or so the theory goes; the reality is somewhat different.

Just how important tourism is, and its tremendous future

potential, is illustrated by recent statistics published by the World Tourist Organisation (WTO), the international body coordinating tourism development. They predict that the current estimate of 415 million international tourist arrivals in 1990 will continue to grow steadily by up to 5 per cent per annum in the 1990s, and this is a conservative estimate. Such an impact will have far-reaching consequences. International tourism currently accounts for 6 per cent of total world exports and 15 per cent of world trade in services, and is forecast to increase by an estimated 4 per cent each year up to the year 2000, by which time it will be unquestionably the world's largest export industry. Total expenditure today on domestic and international tourism represents 12 per cent of the total world Gross National Product (GNP) and the industry employs over 100 million people. In 1988 alone, some one and a half billion domestic and international tourist trips were made involving a third of the world's population. Staggering statistics, I'm sure you'll agree.

Two major factors can account for this extraordinary boom in tourism: firstly, never before have people had so much leisure time nor so much disposable income. Only twenty-five years ago half the British working population had a fortnight or less paid annual holiday. Now everyone is entitled to at least three weeks a year, with four being the norm. With more time in which to travel, people can go further, and this is helped by the second factor contributing to the tourism boom – advances in the field of transport. Widebody jets, express trains and increased car-ownership mean that we can now travel further in less time and at a lower cost. In particular the developments in mass air transport have led directly to the creation of the popular and inexpensive package holiday. In the last five years visits abroad by UK residents have increased by almost 50 per cent.[2]

Inevitably, the increasing numbers of tourists bring corresponding problems. Some argue that tourism creates conditions unfavourable to the growth of the industry itself. As places become popular so they become overcrowded, with tourists often outnumbering locals at peak holiday times. Many small towns and villages are not built to cope with such a great volume of people. Sewage systems get overloaded and the resultant pollution inevitably ends up in the sea; meanwhile the roads and cliffs crumble under the weight of holiday traffic. City officials are even talking about closing off Venice or at best restricting tourist access to certain times of the day. Tomorrow's children can't see the same Europe we did.

Massive demand for accommodation leads to the building of high-rise concrete apartment blocks which are often completely out of character with the resort in which they are situated. Local people soon come to realise that there is money to be made out of tourists and abandon traditional occupations such as agriculture or fishing and move to the resorts where the more lucrative tourist business is found, thus upsetting and distorting traditional patterns of life, a feature common not only overseas but also around British holiday areas. The clash of comparatively rich tourists and their affluent culture with local communities who feel relatively poor often leads to resentment and in some cases an escalating crime rate when it is discovered that there is more than one way of making money out of wealthy tourists. Put yourself in *their* position for a moment. If you were struggling to feed your family and all around you there were rich people spending the equivalent of a week's wages on a round of drinks, how would *you* feel after a while?

Also, local crafts and traditions can become debased as they lose their traditional significance and are abandoned or are churned out *en masse* to fill the souvenir shops, or paraded at tourist folk evenings. There *is* another side to tourism, you see.

The World of Tourists

The unattractive side of tourism is more than just a nuisance: the implications are very serious, especially when you bear in mind that international tourism is still only at the fledgling stage. The British are actually some of the world's most enthusiastic travellers, with 60 per cent or more of the population enjoying at least one holiday a year and with a third of these holidays being spent abroad. Perhaps the fact that we live on an island has something to do with it. At any rate some of our more land-locked European neighbours are much more content to stay at home. Fewer than

50 per cent of Austrians take any holiday at all and only one in five travel abroad. The French take only one-third as many foreign holidays as the British and, surprisingly, only 6 per cent of Americans travel abroad; even then half of these trips are to neighbouring Mexico or Canada. Japan, too, has great potential for a future massive contribution to the tourist population. At the moment there are around nine million Japanese tourist trips abroad each year compared with thirty million British tourist trips. If the Japanese travelled in the same proportions as we do there would be over sixty million trips abroad each year. It is only a matter of time before the Chinese and Eastern Europeans start travelling. Think of the problems then! If we have a tourist problem now, what will happen when these major nations begin to realise their full tourist potential?

Of course it is not simply a question of numbers but more a question of the demands which large numbers of tourists from developed economies in North America, Europe, Japan and Australasia place on locations and countries incapable of sustaining or carrying such a volume. All of us expect a reasonable standard of accommodation: good food, fresh and clean water, plenty of things to do and the opportunity to move around a bit in a quick and comfortable manner. And all this at a reasonably affordable cost. The most economical way of doing this – and very successful it is too in terms of stimulating the business – is to employ the methods common to all big business. Market the product, advertise, create a demand and attract sufficient numbers of people so that the economies of scale make the whole thing reasonably affordable. Aim for quantity and the rest will follow.

The Environmental Issue

The strains of such development can be great. Not only do cultural values become undermined but the physical environment also suffers abuse as short-term economic goals, whether of governments or private investors, override the long-term best interests of air, land and water. Many Mediterranean resorts have such a major sewage disposal problem that beach pollution defies even international sanitation schemes, resulting in incidents such as the 1988 typhoid scare in the Spanish resort of Salou. Constant demand can place an intolerable strain on natural resources if resorts try to develop beyond their environmental carrying capacity. The ever-increasing popularity and availability of skiing holidays has damaged large parts of the Alps through inappropriate development and has exacerbated the damage to Alpine forests attributed to acid rain as trees are cleared to make way for extended ski-runs. The Lake District, one of Britain's most outstanding areas of natural beauty and a tourist attraction that draws many overseas visitors to this country, is now suffering considerable erosion as the soil and natural vegetation is walked away by thousands of feet every year. Such examples underline the complicated nature of many of these issues.

Is all tourist development bad? Should international tourism be restricted? Can we, or should we interfere in areas where people's livelihood is dependent upon tourists and visitors? Is it fair that in order to create a high material standard of living we in the developed world, having exploited our environment and resources, should say to 'underdeveloped' countries such as the Caribbean nations, Thailand and Indonesia, that they should not follow our example but should 'enjoy' only small-scale development without the benefits of large gains in economic growth? The answers are not easy to find and they may change over time. Nevertheless it is essential to seek a balance between enjoyment of the natural and cultural resources of any area, town or country, and the protection and conservation of those very resources for future generations to enjoy.

The other side of the equation contains those positive elements of tourism so eloquently pushed by the proponents of the industry in the early days of mass tourism. Travel 'broadened the mind', 'broke down frontiers' between previously hostile and suspicious neighbouring countries and 'provided jobs and economic growth' in a variety of areas and business sectors. And why not? 'Tourism has become the noblest instrument of this century for achieving international understanding', commented one writer in the early 1960s.[3] A conference on 'Tourism – A Vital Force for Peace' was held in Vancouver in 1988 and led directly to the establishment of the International Institute for Peace through Tourism, which is currently planning the follow-up conference for 1992, entitled 'Sustainable Development in Third World Countries'.

The reality of course lies lost somewhere in the middle, qualified by circumstances and the very real difficulties of weighing up the good and the bad and coming to any positive conclusions. Isn't it ironic that these days people don't even like to be called 'tourists'. 'I'm a traveller, not a tourist', is commonly heard. But we're *all* tourists – one reason we chose this title for the book. So what is the road forward? How can you or I, the 'ordinary tourist', possibly influence such a massive industry? Try going into any large travel agent and tell them you want to do something different from the normal seven or fourteen day package and the chances are you'll have reduced your options enormously.

Environmentally the industry in Britain is changing from being supply-led to demand-led, i.e. responding to and providing for the wants and desires of the tourist. Many of us are only just waking up to the fact that the consumer *does* have influence and he doesn't have to accept only what's on offer. By being selective, stating what we really want, we *can* influence the tourist industry. The proliferation of good, small companies is testimony to the fact that there are choices and operators prepared to listen. The large companies are also not blind to this influence, even if they recognise it initially only as another market niche, and are reviewing their own products. Martin Brackenbury, Deputy Director of the Thomson Holiday group, which currently provides between 20 and 25 per cent of the British overseas holiday market, has said in an interview with us, 'if the consumer wants green packages, he will get them. We want to supply them'.[4]

There are many descriptions of this different approach to tourism: it has been labelled 'green tourism'; 'alternative tourism'; 'responsible tourism'; 'soft tourism'; 'sustainable tourism'; 'Eco-tourism'; 'low-impact tourism'. They all imply a different approach to that of traditional mass tourism. Certainly 'sustainable' tourism appears an appropriate label to hang on such an approach and this term will crop up throughout this book. We have used the term the 'good tourist' to describe our approach.

The Good Tourist

So this book is about being a 'good tourist'. But what does that mean? Although the term rolls easily off the tongue it's a difficult concept to grasp, because there is not one simple easy definition of it. Briefly, 'good tourism' brings benefits to the widest possible number of people, protects and enhances the physical environment and is undertaken in such a way as to sustain these benefits in perpetuity from one generation to the next.

Sustainable development is, if anything, more essential in tourism than in many other economic activities. It's a lesson being hard-learnt in areas such as the Spanish Mediterranean resorts where tourism development has actually spoiled the very things people went there for in the first place. In the long term, tourism and the environment have a common interest in providing thoughtful solutions to avoid such clashes. They should not be thought of as competitive and exclusive alternatives.

When people live in a primarily rural or agricultural society the sustainable development approach is inherent in the husbandry of the resource and intuitive to those dependent on the land. With the breaking of the link with the land in the industrial age and in the developed world, the notion of stewardship from one generation to the next has been down-graded and often ignored. In this respect tourism is no different from other activities in that if you ignore the health of the resource the tourist requires, whether it be sunshine, sand, clean air, attractive surroundings, or cultural experience, then that resource is soon devalued. In the short-term the tourist takes his money elsewhere; in the long run, however, there are only a finite number of places to which the traveller can go. Tourist developers will ultimately have to consider the carrying-capacity of any resort/beach/beauty spot or building to support visitors indefinitely without destroying the attraction. Much of this approach is inherent to the concept of being a 'good tourist'.

To define what makes a 'good tourist' is logical at this stage (though an entire chapter – chapter 6 – is given over to this subject later). The most common simile used in this context is that of the 'host' and 'guest'. Giving the tourist the status of a guest is useful in that it puts the onus for good manners and assimilation with the norm squarely on the tourist. Specifically then, a 'good tourist', like a 'good guest in your home' will:

- respond and adapt to the ways of life and customs of the other environment or country. Put simply, 'when in Rome, do as the Romans do'.
- act in a responsible and sensitive manner towards the people, culture and physical environment.
- not seek to exploit any economic advantage he or she has which diminishes the standing of the host.
- leave any place visited in at least as healthy a state as when he or she first arrived and, if possible when visiting an area of deprivation, leave a practical thank you as a lasting memento of your visit.

Unless tourists behave responsibly and use their enormous economic influence to encourage the tourist industry to do likewise, then in time, inevitably, the quality of the tourist experience will be diminished for all. Ultimately, tourism and the environment should be considered indivisible.

Of course it might be argued that all tourism is bad in that it only squanders resources through encouragement of 'unnecessary' travel and distorts indigenous culture, replacing it with a contemporary 'pop' culture across the world. That is not our view though. Tourism and travel *can* be a force for good if channelled properly, despite the recent negative media reports on the future of tourism, some of which are necessary to stimulate action. Encouragingly, many new types of holidays designed to cater for people with a real interest in their environment have sprung up such as conservation weekends and voluntary service. Their appeal may however be limited. Sustainable tourism is much wider than this – it is an approach, *not a specific product*. It covers such basic ideas as encouraging travellers to be more discerning and aware no matter what type of holiday they choose and to behave in an appropriate manner: for example using local transport rather than their own private car to minimise traffic pollution or avoid holiday honeypots either by choosing a different off-the-beaten-track location or travelling off-season, thereby reducing the traffic, litter and sewage problems inherent in overcrowding.

As well as the environmental issues, good tourists will be concerned for the economic, social and cultural well-being of the country they are visiting. Careful preparation prior to the holiday will make tourists aware of the religious and cultural traditions of the country, taking care not to cause offence by violating these. For example in the Third World photography is often a particularly sensitive area. Indiscriminate camera-clicking can embarrass, frighten or even anger local people who resent the intrusion into their privacy. Clothing, or lack of it, is another bone of contention when insensitive tourists disregard religious conventions and parade scantily-clad bodies around temples, mosques, churches or other holy places. Tourists then become an invasive, insensitive, intrusive force, resented by the locals. Appreciation of these differences makes a holiday more enjoyable for both parties and of greater educational value.

A little bit of forethought also enables 'good tourists' to plan their holiday so that not only they, but also the country they are visiting receive the greatest possible economic benefit. This is especially important when the host country is part of the developing world. By choosing accommodation in a hotel that is privately owned or part of a national rather than an international group, or better still living with locals in Bed & Breakfast; by using local transport and guides; by consuming locally-produced food and drink; by taking home local art and craftwork as souvenirs instead of cheap imported fakes, travellers can ensure that the bulk of their holiday spending ends up in the pockets of the hosts, rather than being 'leaked' back overseas to swell the already comfortable bank balances of the multinationals.

Tourists staying in large multinational corporation hotels, drinking international brands of alcohol or Coca-Cola and eating international cuisine ensure a demand for imported goods which has to be paid for in hard currency. And so the profit which tourism generates to the local economy is often transferred elsewhere. Buying local may sometimes require a little sacrifice in quality but it will be of far more value to the country's economy.

Although the most recent figures show that 80 per cent of tourism takes place in the developed world, the Third World has seen the fastest actual growth in the industry in the last few years. This is well demonstrated by the recent trend towards ever more exotic holiday destinations. As long-haul flights become cheaper and quicker, greater numbers of tourists who are disillusioned with the Mediterranean and overdeveloped resorts go in search of 'something different'. In fact, this search for more in-depth experiences is an optimistic sign since it indicates that people are perhaps more prepared to choose a holiday from which they will learn something, instead of settling for a home-from-home package. However, this type of travel can often result in the greatest clashes, since two completely different cultures are brought face to face. Without care the host nation can be open to exploitation, both by unthinking tourists and foreign companies capitalising on cheap labour and the demand for imported goods to suit Western tastes. This is discussed in greater detail in Chapter 3.

Chapter 2

The Tourist Industry

The Development of Tourism

The phenomenon of large numbers of people travelling relatively long distances for pleasure is really quite modern. Indeed, tourism on a scale large enough to involve the movement of such numbers – mass tourism – developed in two waves. The first started in the late nineteenth century when the freedoms of paid holiday were won by the newly-industrialised working classes. From the odd day out at fashionable seaside resorts this developed into a large-scale leisure industry as people gained more leisure time and had sufficient disposable income to travel, albeit only a fairly short distance from their homes. The second wave began in the 1950s and 1960s with the development of the package holiday where tour operators put together an all-inclusive holiday which, by utilising the economies of scale and the declining real price of air travel, brought the price of a foreign holiday within the range of large numbers of the British public.

On a more basic level, however, man has always been a traveller or explorer. Until recent times journeys were rarely undertaken purely for pleasure. More often travellers were in pursuit of something better than what they had left behind, or they wanted to trade, or else they were off to war or to conquer other lands. The Greeks were probably the first to record trips undertaken for sight-seeing purposes. Travellers such as Herodotus travelled widely and exhibited in their observations of foreign places typical tourist comments. Such individuals were nevertheless extremely rare as the hardships and danger of travel were quite daunting. People also travelled for medical, educational and religious reasons and often included a 'sight-seeing' element in their trip. In Roman times, within a reasonably stable Mediterranean world, the pursuit of pleasure was, of course, very important to the wealthy classes who could afford their own villas in seaside resorts in Greece or

Egypt. With the decline of the Roman Empire and the descent of Europe into a darker age, travel declined, except for emigration, war and conquest and occasionally religious pilgrimage. In the Middle Ages pilgrimages to holy shrines such as Rome, Santiago de Compostela and Canterbury became common, usually made on foot and as part of a large group, sometimes crossing whole continents.

It was probably not until after the Renaissance that travel for pleasure, education and knowledge flourished, if only for those wealthy enough to afford it. The 'Grand Tour' was extremely fashionable amongst the eighteenth and nineteenth-century European aristocracy. This was a trip which could last anything up to three years and which took in all the famous cultural and historic sights of the continent. Records and journals of such trips were published and served to whet the appetite of other potential travellers.

Sir Philip Sidney in 1572 was among the Englishmen who toured Europe. In his essay, 'Of Travel', Francis Bacon encouraged travellers to take letters of introduction in order to meet 'eminent persons of all kind'. Subsequent 'Grand Tourists' such as John Milton and Inigo Jones were deeply affected by the experience, which certainly played no small part in influencing English art, thought and manners of the time. Palladio's influence on Jones has been much-discussed, and the proliferation of Italian statues, busts and paintings in the grand British houses of the eighteenth century prove souvenirs were as important then as they are now. Other notable tourists of their day were Robert Adam, Thomas Jefferson, Boswell, Lord Byron, Wordsworth and Goethe.

The period from 1500 to 1900 also saw the 'Great Voyages of Discovery' to all parts of the world and the first waves of human emigration from the Old World (Europe) to the New World (the Americas and Australasia). Travel became easier and the opportunity to travel more widespread. The 'Grand Tour' became the 'World Tour' and the wealthy of the New World returned to visit the splendours of the Old.

In the late eighteenth century, sea water gained a reputation for its medicinal properties and as a result many small British coastal fishing villages were transformed into fashionable resorts.

Brighton and Weymouth both became popular with genteel holidaymakers after George III paid them a visit to try and cure the fits of madness from which he suffered. The earliest organisation of anything that we would recognise today as tourism started in the mid-nineteenth century. Thomas Cook is historically credited with organising the first ever tourist excursion in 1841 when he sold tickets for a train ride from Leicester to Loughborough as a means of promoting the Temperance cause. Methodist missionary Henry Lunn pioneered the skiing holiday just a few years later.

It was the Industrial Revolution that really began to open up tourism to the working classes. As a result of the widespread social and technological reforms a new middle class grew up, whose increased prosperity meant that they could afford to travel. The Bank Holiday Act of Parliament in 1871, creating four annual public holidays and the Factory Act of 1901, which gave the first ever paid annual holiday allowance of six days, provided the necessary legislation to give the working British public leisure time at no financial loss.

The new railways provided cheap travel to seaside resorts such as Scarborough and Blackpool. Public holidays would see a mass exodus from the large cities of Manchester, Liverpool and Leeds to the coast, for people to be entertained at fun-fairs and shows catering to the tastes of the working man.

People were also becoming aware of a world outside their own direct experience. Soldiers travelling to foreign countries saw opportunities and wanted to return in peacetime. The colonial era brought India, Australia, Africa and other parts of the world into the spotlight of the European colonial powers. The advent of photography provided visual evidence of the existence of the exotic and began to stir interest among the more adventurous to see such sights for themselves. The Taj Mahal, the Pyramids and the Sphinx and the Statue of Liberty are all examples of famous tourist attractions which we travel to see in real life because we've seen pictures of them.

In only a century the holiday has changed beyond recognition. In 1890 a typical family holiday would be a day trip to the nearest seaside town. Armed with bucket and spade and a picnic, the family would take advantage of the half-a-crown cheap-day excursions offered by the private companies who ran the new railway network. The 1990s family, however, is far more likely to take a two-week package tour to the Mediterranean, where the whole holiday will be paid for before leaving home and one can enjoy the guaranteed sunshine.

The Tourism Revolution

Today tourism is not only the world's largest industry but also one of its great boom activities. In the last thirty years the number of international travellers has increased sixfold, an extraordinary rate of growth.

Tourist travel, though, is still very much the privilege of those in the developed world – Europe alone accounts for some sixty per cent of the international travel market, with a predicted increase in international arrivals to over 600 million over the next ten years. Add to that four or five times as many travelling in their own countries as tourists and the total on the move in the year 2000 comes to an estimated three billion.[1] In practical terms these forecasts have horrific implications. Take the Mediterranean coastline, for example. At the moment the resident population of 130 million swells each year by a further 100 million tourists, almost doubling the population for four or five months of the year. United Nations studies reckon that these visitors could number a staggering 760 million by the year 2025 if current growth levels are maintained.[2]

This prediction alone is a worrying one. Until now tourism has remained largely a phenomenon of the developed world.[3] 80 per cent of it comes from only twenty of the world's 233 countries. Moreover, wealthy countries such as Japan and the US have only just begun to contribute to world tourism and their potential for future growth is tremendous. In 1988 the Japanese, with a population of 122 million people, made a total of 8.5 million international trips, compared to the UK figure of nearly 29 million trips out of a population of 57 million. Every year the number of Japanese travelling abroad increases by 15 per cent as Japan's government encourages its people to help counter an enormous international trade imbalance by holidaying overseas.[4] Japan is not the only country whose tourist output is forecast to increase. The number of British overseas holidaymakers has increased by 13 per cent a year and this trend is expected to continue. Changing lifestyles in Eastern Europe, particularly Eastern Germany with its recent access to a capitalist economy, may also play a significant future role in the growth of international tourism.

The Economic Importance of Tourism

The key statistics describing the tourist industry show clearly the global importance of tourism and the impact it has on the world economy. For example, in 1987:

- one out of every sixteen jobs in the world was provided by the industry.
- travel and tourism was the largest provider of jobs in many advanced economies, including the UK, USA, Japan, France and Germany, as well as in developing countries such as Jamaica.
- the industry was an efficient job creator because it was labour intensive. In the US, for example, every $1 billion of gross travel and tourism sales supported more than 5,000 jobs, compared with an equivalent figure of 300 jobs in the chemicals industry for the same outturn,
- travel and tourism sales grossed $2 trillion.
- the industry invested over $275 billion in new facilities and capital equipment and was among the largest purchasers of planes, trains, cars, appliances, buildings and other capital products.

It is not difficult to see why tourism is largely welcomed by countries. According to the *Economist*, if tourism and travel were a country, its Gross National Product (GNP) would rank fifth in the world! International tourist receipts can make a crucial contribution to a country's economy. Receipts from international tourism in 1989 show that the USA was the number one tourism earner, making US $34.3 billion, with France second. Spain and Italy were third and fourth respectively, with the UK taking fifth place with receipts worth some US$11.2 billion. The great tourism powers in terms of revenue are all from the developed economies of the world. However gross revenue is by no means the only, nor indeed the best, indicator of the economic importance of tourism. The contribution of tourism to the GNP of a country gives a far better indication of its importance in relative terms. By this measure tourism is relatively more important to Spain with a 5.2 per cent share of its total GNP, than to Italy which has a larger and more diverse economy with tourism only generating about 1.6 per cent of its GNP, despite the fact that its total receipts are comparable with Spain.[5] Some countries such as Spain, Greece, Portugal and Austria are more dependent upon tourism than others and therefore more susceptible to any disturbance which upsets the tourist equilibrium.

For many Third World countries tourism may be the single biggest source of foreign revenue, a position which can leave such countries dependent on the vagaries of the world tourist market and other extraneous factors beyond their control. For example, tourism is the single largest earner of foreign exchange in India, Thailand, Kenya and the Seychelles. The case of Fiji demonstrates quite clearly the susceptibility of such countries to over-dependence on such a fickle industry. Fiji, prior to the coup of 1987, was dependent on tourism as the major earner of foreign currency. The whole industry was being developed and significant investment was ploughed into the islands. The coup and subsequent political instability eroded the always fragile confidence of tourists to travel and Fiji's economy suffered as a result.[6] The impact of the Gulf War in 1991 on tourism to Jordan, Egypt and Cyprus shows how unstable the industry can be.

On the other hand, while overdependence on any one sector of an economy is always inherently inadvisable, the stark choice faced by many developing countries is tourism or extreme poverty. In the Caribbean, over the last fifteen years, tourism has, in fact, been the *only* industry to show steady growth.

The complex nature of international trade makes the real economic influence of tourism on any one country difficult to evaluate. Put simply, although you may visit any country and spend money there, the amount of money staying in that country may be very small if the goods which you are buying have to be imported or the place where your money is spent is owned by a foreign company. In developed countries this is all part and parcel of the free market and international trade and a reflection of the disciplines of the market place. In a developing country, however, the implications may be more serious. Turning again to Fiji, of the US$70 million gross revenue earned by tourism in 1984 only around 30 per cent stayed in the islands with the rest leaking out either to pay for imported goods or as profits to international companies. In general the more money spent on local products and services the better for the economy of that country. Conversely money spent on imported food, drinks and goods, and foreign-owned accommodation is likely to benefit foreign owners and interests.

In the UK economy, tourism is not only the biggest employer but also the fourth largest export earner, after machinery, chemicals and transport equipment. Earnings from overseas tourism in 1990 totalled $7.5 billion, whilst between 1 and 1.5 million people are employed in the industry. Interestingly, though, the UK is a net loser on the tourism balance sheet – in 1989 expenditure by British tourists abroad exceeded tourism income from international tourist arrivals by $2.4 billion. In fact this negative tourism account balance has grown by 120 per cent over the last six years. By contrast the USA, the world's largest economy, has shown a remarkable increase in tourism receipts to become the world's number one international tourism earner. The cliché of the American tourist abroad spending vast amounts of dollars is now being replaced by that of increasingly wealthy Europeans and Japanese travelling to the States.

Why be a Tourist?

Why do we become tourists and why has tourism grown so much? The major reasons for the development of tourism are the increased amount of leisure time, the increasing ease of long-distance travel, more disposable income and the opportunities seized by entrepreneurs in the leisure and tourist industry to utilise these factors. So whilst many of us do genuinely believe that we are travelling to relax and get away from it all the tourism, travel and leisure business could be seen as just one more aspect of the highly organised, urbanised and industrialised lives that most of us lead. The travel agent is often doing no more than selling us the dreams with which we reward ourselves for the stressed lifestyle we find ourselves living and perpetuating. We are simply consumers of another product – holidays.

It is fashionable to perceive holidays as rare chances to get away from it all, yet much of the excitement of going on holiday is in the anticipation and preparation.

To a certain extent tourists can be selfish, choosing to do only the things they like best because they have paid to enjoy themselves. Holidays give us the chance to relax and unwind, a vital recuperation that enables us, we feel, to face another year of work. Many people, of course, travel with a particular purpose in mind. Often this goal is a change of climate and to enjoy a destination that will allow us to do our relaxing in warm sunny weather. Another common goal is the pursuit of a particular hobby or activity such as skiing, walking, fishing or golfing. Less often we travel in order to learn something new or to broaden our horizons. These holidays need not be to far-flung exotic destinations but could be to a European neighbour where we intend to acquire some proficiency in a foreign language or to learn about the history and culture of our holiday destination.

Mass Tourism

The current large-scale and highly developed nature of the tourist industry is often called 'mass tourism'. Development on such a scale has provided relatively inexpensive holidays to large numbers of people. Certain resorts were chosen by investors and business interests to develop as tourist destinations and attractions. Tourists were then encouraged to travel to such resorts by the offer of cheap travel and accommodation i.e. the 'package holiday', whereby the client received an all-inclusive product with minimal hassle. The most obvious example of this has been the Spanish Costas and island resorts such as Majorca, where British tourist numbers increased dramatically in the 1960s and 1970s. The impact on these destinations was quite incredible and changed them out of all recognition. In only a few years they have had to come to terms with their participation in an industry that has mushroomed and this change has often found them unprepared for the long-term effects; the scale of change has been staggering.

Such places as Benidorm, Palma, and Torremolinos bear witness to the change. Torremolinos, for example, was a quiet, poor, fishing village in the 1950s. Originally 'discovered' by the wealthy 'society' it quickly grew into a fashionable resort, attracting 'le beau monde'. Fashion being forever fickle, the wealthy deserted Torremolinos for pastures new and it went on to develop into the archetypal 'concrete Costa' catering for ever-increasing numbers of northern European package tourists in cheap hotels and apartments.

Even relatively distant underdeveloped countries have experienced enormous increases. In 1969 1,800 people visited the island of Mauritius in the Indian Ocean. Only twenty years later around 180,000 tourists visited the island in a single year and this number is expected to increase to 300,000 by the end of the century.[7] One of the greatest social consequences of mass tourism is the negative reaction it causes in the host populations. Tourism in bulk has an inhuman, impersonal face, and many local people resent what they now see as an annual invasion by the tourists. Most societies have a tradition of hospitality towards visitors. In the early days of travel and tourism the small scale of visiting parties ensured that these traditions often served a destination well in welcoming visitors and encouraging a reputation of friendliness. Once large numbers arrive and keep on arriving over a season the initial friendliness and hospitality becomes strained. Those locals who derive no direct benefit from tourists then start to feel resentful of all the disadvantages of tourism – congestion, noise, rowdiness and general inconvenience. Younger generations who have grown up with mass tourism feel a much greater resentment at the appropriation of their country, region, town or whatever it is that seems to be taken for granted and unappreciated. For the tourists themselves each experience is a new one and unique but the hosts must endlessly repeat their hospitality and it is little wonder that their friendly smiles can become somewhat fixed.

Tourist Destinations for Tourists from the UK

In 1951 of the twenty-six and a half million holidays taken by the British, twenty-five million were spent in the UK while one and a half million ventured abroad. By 1960 with the introduction in 1950 of the inclusive or package tour, holidays abroad accounted for three and a half million of the total thirty-five million holidays. The most dramatic growth has been the rapid rise in popularity of the package holiday which now accounts for two-thirds of all holidays taken abroad and which has been responsible for luring a third of all British holidaymakers away from a UK holiday.

Recent trends in the foreign destinations visited by Britons make interesting reading. In 1989 of the thirty-one million foreign trips undertaken, the percentage breakdown by destination was:

Geographic area	percentage
EEC countries	72
Other Western European countries	12
Eastern Europe	1
North America	7
Australia and New Zealand	1
Rest of the World	7
Total	100

The most popular countries were France (21 per cent of the visits abroad); Spain (20 per cent); Ireland (7 per cent); US (6 per cent); Greece (5 per cent). Over the last three years whilst overall travel has increased by 12 per cent, some areas have done better than others, reflecting the changing patterns in travel and tourist destinations. North America as a destination increased by 41 per cent, and France by 22 per cent, whilst trips to predominantly Third World countries increased by 18 per cent. Conversely Spain, Greece and Yugoslavia have seen numbers fall in absolute terms, by 6 per cent, 12 per cent and 14 per cent respectively, all during a period when numbers of people travelling are increasing.[8]

There is thus a discernible trend away from the traditional

package destinations of Spain and Greece to countries offering something different to the normal package such as France, Ireland and Germany, or to farther flung and hence more exotic destinations.

So, an increasing number of UK tourists are choosing to spend their holidays in the less developed countries of the world encouraged by cheaper airfares, new inclusive packages and disillusionment with the traditional Mediterranean resorts, not to mention the lure of advertisements describing 'exotic' destinations in quite outrageous terms. These far-off countries only account for a small proportion of tourist activity but currently they represent one of the fastest growing sectors of the market and one where the impact of tourism might be the greatest. Historical colonial connections, a common language and the chance to sample the exotic are all reasons which draw people to countries such as Kenya, the Seychelles and North Africa, particularly Tunisia and Morocco. Many offer a sunny climate during the northern winter while the natural attractions of countries like Kenya and Zimbabwe in terms of wildlife and national game reserves, draw visitors in search of adventure. In fact Zimbabwe has directly profited as a result of the voluntary ban by almost all the major British tour operators in tourist trips to South Africa and has now established itself as a viable alternative for safari tourism.

Asia, in particular, is experiencing a boom in tourist arrivals, especially from the UK. India, with its strong colonial connections, relies heavily on the British tourist market which contributes around 15 per cent of its total tourist arrivals. To encourage this further the Indian government invested £11.5 million in the development of 'beach tourism', largely centred on the beach resort of Goa, though this was not without continuing controversy and opposition from locals.[9]

The Caribbean is another popular holiday destination for the British who, together with the French and the Dutch, accounted for 74 per cent of all European visitors in 1985. Colonial ties are obviously a strong attraction with 80 per cent of British tourists visiting former or present British colonies such as Jamaica, Barbados, the Bahamas and St Lucia.

Not surprisingly such a large movement of people on a regular basis to recognised holiday resorts or regions has resulted in many of those areas being changed permanently and not all in a positive way. To cater for large numbers of people the infrastructure of a resort and its support facilities must undergo intensive development so that it is able to meet their daily requirements. The destination thus often becomes similar in many ways to what the tourist left at home, the proverbial 'home-from-home', where almost the only remaining differences are location and climate. Chapter 3 explores these issues in greater detail.

Travel Agents and Tour Operators

During the period of growth of foreign holidays there was also, not surprisingly, a corresponding increase in the number of travel agencies and tour operators in the UK. Such people have a critical role in the travel industry as they can very much influence the type of holiday available and the promotion and marketing of particular destinations. From the tourist's point of view a travel agent should be able to find out the answers to any questions the prospective tourist might have and advise on the destination and associated facilities. Unfortunately wages in the travel trade are not good and the quality and training of staff is often not up to the level of demand. Sales assistants often know little more than is in the travel brochures and either don't know how to find out information or consider it not worth their while to do so. Accordingly, if you find a good agent hang onto them and use them!

Travel agents earn their living by getting commission from selling holidays on behalf of the tour operators. Tour operators are the people who put together holidays and tours by negotiating deals with airlines and hotels, and other associated elements of the typical package holiday. Because they operate on a large scale their bargaining power is strong and they can get prices and deals for all different components of a package cheaper than can any individual traveller or tourist.

It would be wrong to forget that travel agents and tour operators are commercial businesses. Like all businesses they require to make money in order to continue to provide the service – holidays

– that most of us want. There is nothing unusual or wrong in this. It is how most economic activity in the Western world takes place. The current international movement towards market-place economies will reinforce this approach. Unfortunately, in the tourist industry it is not easy to make the link between those who make money out of tourism and the true cost of providing those features which attract tourists in the first place – the sea, the beaches, the mountains, the landscape, the countryside and so on. There is a growing awareness now of the need to make that link and increase the responsibility and accountability of the industry. This makes sound economic sense in the long term also – it is in no one's interest to downgrade the very places that attract visitors in the first instance.

The events of the Gulf War of 1991, on top of the recession hitting the UK market at the same time, illustrate clearly the fragility and low margins of those in the tourism and travel business. As far as many smaller operators are concerned, it is debatable whether they are in the business to earn a living, or simply because they enjoy the involvement and have a tremendous enthusiasm for the business, which helps them overcome the various crises that periodically hit the market. The Civil Aviation Authority predicted in May 1991 that there would be a 15 per cent drop in revenue in 1991–2 as a result of the slump in package holiday sales, figures which follow on from the 2 per cent drop in holidaymakers going abroad in 1990, compared to 1989. All this has to be put in perspective, of course. The long-term trends still strongly indicate a continuing increase in tourism and travel. All that has happened in Britain is that the tourists who didn't go abroad stayed within the UK. Because travel and tourism are considered by many as 'luxury goods', the cutbacks tend to be greater in bad times, while conversely the increase is more rapid in good times.

The 1990s have seen a shake-up in the UK travel industry. The demise of the International Leisure Group has altered the structure of the whole sector. However, further upheavals are likely as the government unveils its response to an EEC directive on package travel which will come into effect in 1993.

The big five travel agents are Lunn Poly (500 high street

shops); Pickfords Travel (332 shops and 75 business travel outlets); Thomas Cook (330 shops and 64 business travel outlets); A.T. Mays (297 shops), and Hogg Robinson (218 shops and 42 business travel outlets).

For tour operators, the market is dominated by the Thomson Group (3 million holidays), Owners Abroad (760,000 holidays), and Airtours (1.5 million holidays), who between them sell 50 to 60 per cent of all package holidays in the UK. These operators are followed by Best Travel (300,000).

The Association of British Travel Agents (ABTA) is the organisation which represents 90 per cent of all tour operators and travel agents in Britain. Because its principal duties are to create as 'favourable a business environment as possible' for its members, it has been criticised for not taking sufficient account of the impact of tourism on holiday destinations. It is a self-regulatory body but recently has been subject to much criticism within the trade, and has undergone an internal restructuring in the light of the travel trade's recession difficulties. This is unfortunate as there is undoubtedly a role for ABTA to play. There already exist codes of conduct for travel agents, whilst tour operators are required to provide bonds to protect customers in the event of financial failure. It would be heartening to see the industry itself develop a code of conduct in terms of the social and environmental impact of tourism and seek to disseminate knowledge and better attitudes towards such issues both within the industry and outwardly to customers – the prospective tourists.

The influence of tour operators on the type of holidays available and the destinations on offer is enormous. It had been believed by many travel industry observers that as the travel industry developed and matured there would be an increase in independent holidays abroad: in fact the evidence in recent years points to an increasing proportion of package compared to independent holidays. Something like two-thirds of all foreign holidays are packaged. The package market has become increasingly sophisticated and now everything from a trip to the Arctic wilderness to a tour of the Pyramids in Egypt or a Caribbean cruise are all available from a brochure, all ready-made.

Advertising plays a key role in promoting holidays and trips.

Many of us do choose our holiday destination based on the advertising 'blurb' in brochures. The selling of images and dreams is now highly subtle and extremely influential. A recent WWF study in Germany investigated 107 travel catalogues from 46 German travel promoters and organisers, from the point of view of the information given to prospective clients on the true nature and environment of the potential holiday destinations. In many cases the information was little more than an advertisement. Kenya, for example, was reduced to beaches and game parks; the Maldives to beaches, palm trees and a brightly coloured underwater world; and Venezuela to beaches and lush green jungle. No mention was made of the need to help conserve these wonderful assets, nor indeed of the associated local culture, thus reducing the destinations to simple tourist attractions without the perspective of how these attractions relate to the rest of the country.

Clearly travel companies have a moral responsibility to market their products honestly and the evasion of this responsibility using the excuse that they are only providing what the customer wants is common place and obviously wrong. Things are beginning to change as advertising standards are improved but it is still vital that we should be aware of the influence of advertising and its potential to distort reality. It can lead to false expectations of a country or destination so that a visitor will be disappointed on arrival to find things not as anticipated and feel cheated by the tour operator, or even the country itself.

Chapter 3

The Negative Impact of Tourism

As you lie there in the sun, soaking up the rays and enjoying a quiet drink by the pool or on the beach, surely all is right with the world? If only everyone could experience a few weeks away each year we'd all be the better for it . . .

Though there has always been a certain snobbish view taken of mere 'tourists' and the trail of destruction they leave behind them, the criticisms aired in the past have today grown into major doubts about the future of tourism. Why has this situation occurred? Why is tourism, seen by so many people for so long as the economic 'miracle' solution for many impoverished regions and countries, now being questioned and viewed in a new light? It brought jobs, foreign exchange, helped to develop economies and, with its service industry 'pollution-free' image, made a major contribution to international harmony and understanding. Or at least that was the way it was sold to bewildered locals in places such as Greece and Spain, who saw their sleepy coastal villages transformed into pulsating, thriving resorts in a very short period of time.

Briefly, the answer to these questions is sheer numbers and the fact that tourism as an industry is not divorced from other forms of economic activity in the late-twentieth-century developed world. The enormous volume of tourists and the corresponding scale of development necessary to support their rather peripatetic existence is the real reason for the problem. One tourist in a Greek village of 200 people is a guest who can be easily accommodated, fed and introduced to local society without having any great impact on the local environment or culture. Introduce twenty a week and the strain will begin to tell, as demands increase and the impact of outsiders affects local attitudes and lifestyles. Expand this to 1,000 every week and you have a problem – the village is swamped. Every one of those 1,000 will want accommodation and services at least as good as those at home. They will want to be entertained, to travel and move around; in short to impose their standards and expectations on the place in which they've come to holiday. Inevitably, all this demand requires investment in order to provide the necessary facilities which will include hotels, shops, restaurants, roads and access to recreation areas such as beaches and lakes. Local entrepreneurs will benefit by responding to these demands, but before long the local fisherman or farmer has trouble finding labour, as workers desert to more lucrative employment in the tourism business. Outside investors see opportunities that local people can't afford to exploit, so they move in on a resort and it develops without local control. This scenario has been repeated time and again around the Mediterranean.

We shouldn't really be surprised by all this. Tourism increases people's consumption by stimulating more economic activity. That's fine when this activity produces development which is beneficial to the local community. Sheer numbers, however, can very quickly overload both natural and man-made systems: delays at the airport and raw sewage on the beaches are both symptoms of the same problem; the carrying capacity of the system has been exceeded. The great environmental debate currently under way does not consider tourism in isolation. Quite the contrary. By becoming a tourist most of us increase our consumption of resources and put further pressure on particular environments. Tourism encourages us to travel more, contributes to a greater demand for natural resources and raw materials – paper and packaging, heating, air-conditioning, water – and stimulates the development of previously virgin natural habitats such as beaches, coastlines – imagine the Spanish Mediterranean coastline forty years ago – and mountain tops. Perhaps most importantly, it puts many people into natural environments which tend to be more fragile than man-made ones. The simple mass of people and their associated demands and activities can lead to highly stressed and overloaded natural environments. Man-made environments are much easier to restore and manage than fragile natural areas. Thus cities and purpose-built resorts are geared up to coping with large numbers of people, or at least have the potential to do so in a way that minimises environmental damage. There is, however, still a danger of overloading even these tourist areas. Overuse tends to degrade and devalue, causing one of the classic tourist dilemmas of constantly searching for somewhere new or different. Man-made attractions often reach a peak and then decline due to more effective competition or insufficient investment in modernisation. Undoubtedly one of the major requirements in striking the balance between tourism and environmental preservation is to reach a fair and happy medium between demand and overuse.

The Different Types of Impact

Basically tourism can impact on an area or country in any one of three broad categories: economic; cultural; environmental. They are not by any means mutually exclusive. Indeed there is a considerable overlap between all three. The three categories, however, serve as a useful basis for analysis.

(A) THE ECONOMIC IMPACT

'Tourism boost to the local economy' is a newspaper headline which is familiar to many of us as we read about another proposed tourist development either at home or abroad. Local politicians mouth the usual platitudes about the benefits of such tourism, and in general everyone appears happy. Under closer scrutiny it is apparent that very rarely have the costs and the benefits for such developments been properly weighed and considered. Undoubtedly there are economic benefits to be gained from tourist development (and we shall explore them in Chapter 4) but there *is* another side. International tourism involves people travelling from one country to another. It follows, therefore, that there will be a balance of income from foreign tourists to be set against expenditure by tourists of that same country, when *they* go abroad. In this respect tourism is no different from any other internationally-traded commodity or service. If more money is leaving the country than is coming in, then international tourism may, on balance, be resulting in a net transfer of wealth from that country. Britain is a good example of this. In 1988 and 1989

British tourists spent over £2 billion more each year on overseas travel and tourism than foreign visitors spent in the UK. And current trends show the gap to be widening.[1]

Job creation is another area which brings uncertain benefits. Whilst jobs are undoubtedly created – there has been an increase of 19 per cent in jobs directly related to tourism in the last decade and this now accounts for 6 per cent of all UK employment – many of these are unskilled or casual, entailing lower wages and employment for only part of the year. In some situations such jobs can upset the local labour market by taking people away from traditional occupations such as fishing or farming. In Third World countries this can be a destabilising factor, even if many of us can sympathise with and understand the wish to change from the hard physical labour of working the land to the 'glamorous' lure of tourist jobs and their association with the leisured classes.

Many poorer countries see tourism as a means of earning foreign currency. Not unnaturally if they are poor in other natural resources then they will seek to exploit those things which they might have in abundance – a good climate, beaches, clean warm seas. Nevertheless, much of the revenue earned by such countries can actually 'leak' back out of the country to pay for imported goods and services such as food, drink, building materials and cars, not to mention the employment of expatriate 'experts' and managers who are brought in to help run the tourist developments. Such 'leakage' occurs in virtually all tourist destination countries but has a most marked effect on the undeveloped countries of the Third World. As much as one-third of the food bill of most luxury hotels in Jamaica accounts for imported food. On package tours to the Third World as little as 22 per cent of the retail tourist price will remain within the economy of the destination country. Foreign ownership of such essential tourist-related businesses as airlines and hotels is another major contributor to this leakage.[2]

There have been many studies of the effect of leakage of revenue. In cases where a package holiday includes a foreign airline but all other components such as accommodation are locally owned, only 40 to 50 per cent of the holiday retail price stays in the host country. If, however, both airline and hotel are foreign owned this percentage drops dramatically to 20 or 25 per cent.[3] In the most extreme cases an incredible 80 to 90 per cent of foreign tourist money leaks straight back out of the country, having done the host population almost no economic good whatsoever. The Pacific Islands are a good example of this. In 1984 over US$70 million was generated in Fiji by its hotels alone, this sum being about half the total national revenue for that year. In the same year tourism was responsible for 39 per cent of all sales in Guam, and Hawaii earned US$3.6 billion through tourism in 1985. However, a study in 1987 estimated that of the total income from tourism in the Pacific Islands, only 25 to 30 per cent stayed there. A staggering two-thirds of tourist money went to the multinational companies who control the industry.[4]

For wealthy industrialised countries which already have well-developed transport, sewage and other essential infrastructure systems, the capital costs associated with tourist developments will be less, and the profits accordingly greater. For a poorer country with limited fresh and running water, a poorly-developed road network and no efficient sewage system the costs of tourism development may be considerable.

Economic development can bring with it further problems such as the escalation of land and property prices, with a consequent impact on the local population who are unable to compete with a combination of wealthy developers, tourists and second-home owners. Tourist resorts within reach of urban areas throughout Europe such as the Scottish Highlands, Wales, the South of France, the Algarve, Corsica and the Spanish Mediterranean coast have all suffered to some extent from this. Local feelings of resentment in such situations can run deep as evidenced by recent nationalist arson attacks on tourist or second houses in Wales and Corsica. The swing to a tourist economy can also result in the pricing of goods and services above the reach of the permanent residents. In extreme cases whole economies can become dependent on tourism with disastrous results, when events like the Gulf War can overnight ruin an area's tourist industry.

Countries such as the Seychelles with 57% of total foreign exchange dependent on tourism-related activities, Jamaica (53%), Nepal (20%) and Kenya (17%) rely heavily on tourism.[5] Such over-dependence can result in economic problems should there be a reduction in tourism earnings due to such diverse factors as political unrest or tourist fashion – is a country 'in' or 'out' this year? Not only Third World but also developed countries such as Greece and Spain have a heavy reliance on tourism. Is that wise when dealing with such a fickle business?

The seasonality of much tourism activity places a strain on both the natural environment and the local infrastructure due to the necessity of cramming as many people as possible into resorts, airports, beaches and so on. Seasonal employment for the local workforce poses particular problems in areas where tourism is the main activity. Hotels and other tourist-related facilities close for the rest of the year, representing an under-utilisation of the resource and associated workforce. Of course the current system whereby school and workplace holidays are concentrated into July and August, exacerbates the whole situation.

(B) THE CULTURAL IMPACT

International tourism leads to the random intermingling of tourists from different races, countries and cultures which in turn leads to interaction between people out of their normal social context, with all sorts of opportunities for misunderstanding. While tourism is very often put forward as a vehicle for increasing international understanding, more commonly it seems that visiting other countries only serves to reinforce the stereotyped images and prejudices many of us feel towards other nationalities. It seems unfortunately true that tourists from one country, when abroad in large numbers, do seem to act according to some perceived national identity – witness the English 'lager-louts', the 'chaotic' Italians, the 'rude' French, the 'loud' Americans and the 'arrogant' Germans. We all suffer from such misconceptions! Nevertheless it is not easy to separate the behaviour of people as tourists from the rapidly changing normal lifestyles of most Westerners. Undoubtedly cultural impacts are closely related to the sheer numbers of people involved as tourists. Overcrowding in resorts puts pressure on both the tourist and host population.

In Hawaii, Japanese and American tourists out-number Hawaiians by five to one in the summer; tourist buses are regularly stoned by locals shouting 'tourists, go home!' In places where tourism becomes the major employer traditional activities are abandoned in favour of the greater material opportunities and 'bright lights' on offer.[6]

Inevitably, constant contact with alien customs and cultures results in a weakening of native traditions and values. Young people find the wealth and pleasure-oriented lifestyle of the tourists enviable and try to copy it, not realising that the way tourists conduct themselves whilst on holiday is not necessarily the way they behave when at home. In very poor countries often the only way to possess the wealth of the tourists is to steal it from them and many of those who gravitate towards the tourist resorts in the vain hope of employment turn to crime to provide an income. Theft is not the only form of crime that makes money out of tourists: prostitution or 'sex tourism' is a way of life in many parts of the world, especially in Bangkok and the Philippines. It regularly involves girls as young as thirteen who send their earnings home to their villages to support their families. They can nearly always make more money by selling their bodies than by legal employment and in this way rural life breaks down. Tour operators, notably in Germany and the UK, actually sell sex tours as just another travel product. We met a German tour operator in Bangkok who laughingly referred to his holiday as 'the ultimate in activity holidays'!

It can be argued that tourism is a positive force in sustaining local crafts and traditions that would otherwise have been lost through lack of interest or relevance, and this is true in very many cases. There are, however, an equal number of examples of the cheapening or 'prostitution' of cultural traditions through their development as a form of tourist amusement: as an ex-President of the Hawaiian Visitors Bureau confessed 'Since real cultural events do not always occur on schedule, we invent pseudo-events for the tour operators who must have a dance of the vestal virgins precisely at ten am every Wednesday.'[6] Visitors to Tunisia are encouraged to attend a 'traditional' Bedouin feast. This custom, practised over centuries, involves elaborate camp-fire feats with singing and dancing that embodies deep religious and symbolic meaning for the Bedouins. The version to which the tourists are taken in bus-loads is a carefully stage-managed performance where wine is distributed liberally to the audience and visitor-participation in the belly dancing show is encouraged. The whole experience, although it claims to do so, gives no real insight into Bedouin culture.

When tourism was still in its infancy, foreign visitors were often overwhelmed with free food and accommodation and personal invitations to events such as family weddings etc. In some parts of Greece this xenophilia – love of foreigners – is still in evidence, but elsewhere the huge numbers of tourists make this charitable practice impossible; it risks bankrupting the hosts and the relationship between host and guest becomes purely a commercial one. In Third World countries tourists are often shocked at the materialistic attitude of the locals who will charge to have their photograph taken and whose English vocabulary seems to consist solely of the words 'How much?' Yet these people see the tourists as wealthy and think it natural to expect them to pay for services rendered.

The international culture that is now becoming universal, thanks to the 'benefits' of TV, Coca-Cola and multinational corporations, is spreading inexorably, via tourism, into previously remote and isolated places. Whether this would happen anyway is a moot point but you might argue that tourism itself is quickening the pace of change to this international culture.

(C) THE ENVIRONMENTAL IMPACT

The phrase most often used by tour operators and holidaymakers alike when describing an attractive resort is the word 'unspoilt'. We all want somewhere untouched by man, it seems. So the environment is often the single most important factor in deciding if a resort is attractive or not. Ironically though, as we have seen, without careful pre-planning and control tourist development can destroy this very resource. And there's only so much of the world left that is genuinely 'unspoilt'.

There are many ways in which the environment has already suffered at the hands of tourism:

The Negative Impact of Tourism

- inappropriate development: the urban sprawl along the Mediterranean coastline, stretching almost unbroken from Spain into France, Italy, Greece and now to Turkey, is designed principally to cater for a large influx of people at certain times of the year. Much of this development has been done with minimal regard for the impact on the environment and the need for adequate support services such as roads, water and sewage.
- loss of natural habitat: beaches and coastal areas are prime targets for developers. A fragile ecosystem such as the Alps is now subject to 40,000 kilometres of ski-runs and 14,000 ski-lifts. The scars of the intensive use are apparent in the loss of forest cover and eroded soils so evident in the summer months. Tourism is the major reason for this development and its clash with the natural environment is all the more ironic in that 'unspoilt' or 'natural' sites are most often chosen as the site of new tourist developments. In Spain even the superb Cota Donana National Park, the last great lowland wilderness sanctuary in Southern Europe, is currently under threat from tourist development proposals. The destruction of the coastal habitat in the Mediterranean has placed such species as the Mediterranean monk seal and loggerhead turtle on the verge of extinction. Even safari tourism, which has done much to engender a motive for conservation in many African countries, has been allowed to run out of control in many national parks; many irresponsible safari operators follow game across country in vehicles, driving big game species to the peripheries and killing vegetation, which leads to soil erosion.
- extinction of species: wildlife can be affected directly through hunting and cropping to provide souvenirs for tourists. Snakeskin and alligator belts, handbags and shoes in Mexico, black coral in the Caribbean, elephant ivory in the Far East, turtleshell throughout the tropical and sub-tropical world, all indicate the extent of the trade. The Convention of International Trade in Endangered Species of Wild Fauna and Flora (CITES),

to which the UK is a signatory, controls trade in rare wildlife, but much of the trade flourishes through ignorance on the part of tourists buying such souvenirs. Live animals are often used for posing with tourists. The problem has been highlighted in Spain where chimpanzees and lion and tiger cubs are taken from the wild, sedated and used for photography, and then killed once too large to be handled safely. Such degradations have a serious impact on wild animal populations.

- pollution of the seas and water courses: particularly raw sewage and industrial pollutants which threaten the quality of beaches as both a tourist facility and a healthy habitat.
- disease: susceptibility to disease increases as tourists travel further to more exotic locations, and encounter conditions and illnesses alien to them. Whilst extraordinary advances in medicine have been made there is some evidence that diseases such as malaria are increasing again due partly to the greater mobility of people who often adopt the 'it'll never-happen-to-me' approach with a consequent lack of precautions.
- loss of the spirit of a particular place: whether it be a wilderness area or a famous building or a monument such as Stonehenge or Westminster Abbey. These attractions have a very definite 'spirit' which is a large part of their attraction to visitors. Overcrowding in such places quite simply destroys the very reason why many people went there in the first place.

It is apparent that the economic, cultural and environmental impact of tourism are closely intertwined. You cannot develop a Mediterranean coastal village or sleepy Caribbean island without altering it to some degree. Any negative impact on the environment is bound to reverberate at some future stage. This is the essence of many of the problems currently being experienced by the Spanish tourist industry, where unstructured and unsympathetic development has reaped a short-term economic gain but at great cost to the quality of the man-made and natural environment. That such an approach is unsustainable is witnessed by the desertion of tourists to other countries and destinations. Commercially too, sustainable tourism makes sense!

Two areas in particular illustrate well the impact of the tourist; the Mediterranean and the Alps.

THE MEDITERRANEAN

The Mediterranean, undoubtedly, has been the centre of international tourism as far as Europe is concerned. The trend has been for a large influx of Northern Europeans to move to the shores of the Med in summer. Attracting over a third of all international tourism, the sea and its surrounding coastal areas have borne the brunt of the early package-holiday boom and are now counting the environmental cost for such popularity.

The shoreline comprises eighteen countries, speaking ten official languages and practising three of the world's five major religions. With its historical and cultural richness and its outstanding natural beauty, it is little wonder that it attracts the tourists. But now that the effects of the last twenty to thirty years are telling, the future is looking ominous. In 1990 mainland Spain faced a 50 per cent slump in bookings, while the package market saw an overall drop of 20 per cent.

Tourism in Greece has been declining since 1981, despite its undoubted cultural and climatic attractions, due to uncoordinated planning and a deterioration in the quality of tourist services. Customers are fed up with the 'concrete Costas' and holiday hassles such as long flight delays, ugly and sometimes structurally unsafe hotels overlooking building sites, dirty beaches, sea water with algae and outbreaks of typhoid and other unpleasant diseases.

'Undiscovered' and 'unspoilt' destinations ripe for development have risen and fallen as new ones catch on. The beaches of Spain, Portugal, Italy and Greece have taken it in turn to be the 'in place'. Most recently it's been Turkey's turn.

The problems of the Med are a microcosm of what's happening around the world. Traffic congestion is a problem, as tourists in coaches, caravans and cars squeeze on to roads that were not constructed to bear so vast a load; frequent fatal accidents involving tourists as both pedestrians and drivers attest to the total inadequacy of the transport system which was never designed with the needs of tourism in mind but merely adapted from the old narrow coastal routes serving the small fishing communities. Traffic congestion is not restricted to the roads. Increased air traffic puts intolerable strain on already overloaded facilities. Passengers have recently suffered delays of forty-eight hours or more at the height of the season. Industrial disputes compound the problem as airports find themselves running short of runways to cope with demand. During its peak period of July and August, Heathrow has flights taking off and landing every minute. A fifth major British airport has been started to spread the load. Overcrowding of the sky results in more fuel pollution of the air, contributing further to the decay of natural habitat and historic monuments.

Pollution of the natural environment is, in short, killing the Mediterranean. Geographically the Mediterranean is more like a huge lake than a sea; with over a million square metres of water and a tidal range of less than a metre, the water is renewed just once a century. Six major rivers empty into the Mediterranean: the Ebro, Rhone, Po, Arno, Tiber and Nile. Large-scale chemical and metal processing plants situated on these rivers are responsible for the dumping of industrial pollutants such as the deadly poisonous mercury and high levels of this metal have been found in mussels, fish and fishermen. The report of a Sicilian government enquiry conducted in the industrial port of Augusta revealed a long-term public health crisis which had been ignored for decades and which was attributed to the unregulated pollution emissions from seven nearby petrochemical processing plants, three of which have now been closed until they can be made safer.

France, Spain and Italy have nuclear power stations which release radioactive waste directly into the sea and it is calculated that about half the world's oil pollution occurs in the Mediterranean. Oil tankers visiting the thirty or more major refineries (dealing with over 100,000 barrels a day) along the coastline flush residual oil out of their tanks into the sea; much of this washes up on nearby tourist beaches.

Agricultural waste is another major culprit. In 1985 115,000 tonnes of phosphates and 340,000 tonnes of nitrates were released into the water off the French coast. These pollutants are often swept into the Mediterranean, via the rivers, from their run-off point far inland. This causes a build-up of polluted water around the river estuaries.[7] Similarly raw sewage is piped relentlessly into the rivers and sea, largely because the influx of visitors overloads sewage disposal systems in the major resorts and cities. Worst affected areas are along the Spanish and Venetian coasts and the west coast of Italy. Sewage has contributed to outbreaks of dangerous diseases such as typhoid, cholera, hepatitis and dysentery in Italian and Spanish resorts. It is sewage and agricultural waste that are largely responsible for the red algae 'blooms' that have hit the beaches and the headlines over the last couple of summers. The algae arise from natural causes but the enormous blooms which left a layer of slimy mucus on seabed and beaches in the Adriatic last year were a direct result of the 'enrichment' of seawater by the nutrients contained in untreated sewage and other pollutants. The dirty waters of the Po are the major culprit, and the cleaning-up of the Po basin would make a significant contribution to the fight against further algal blooms. Known as 'red tides', these carpets of algae coat the seabed in rotting, poisonous slime and a surface layer starves the water and its inhabitants of oxygen. In 1987 a bloom off Italy's Adriatic coast killed all marine life within an area of 400 square miles south of the mouth of the River Po, below Venice. Two years later this area again suffered the same fate and this time the disastrous consequences for its economy was a total loss of US$3 billion in tourist and fishing revenues.[8]

The Mediterranean is now the dirtiest sea in the world and its wildlife is suffering as a consequence. Fish breeding grounds are affected and fishermen now struggle to make a living out of their traditional occupation. In particular there has been considerable publicity over the threat from tourist development to two species of marine wildlife: the monk seal and the loggerhead turtle. The rare Mediterranean monk seal, one of the most endangered species in the world, is threatened with extinction because its breeding grounds have been developed as tourist resorts. Only 350 of the species, mainly living outside the Mediterranean, were known to be alive in 1986. The survival of the loggerhead sea turtle is also threatened by the tourist industry as its beach nesting sites are bought up by the developers. The development of undisturbed beaches in Spain, Corsica and southern France has led to the disappearance of turtles from these countries and the conservationists' spotlight has recently been pointed at Greece and Turkey as final refuges for a vanishing breed. International pressure and World Wide Fund for Nature publicity, distributed by sympathetic tour operators such as Thomson, Intasun and Sunmed, highlighted the turtles' plight, resulting in a 1987 law passed by the Greek Government protecting their turtle population. Laganas Bay on the Greek island of Zakynthos is now one of the major turtle nesting areas in the Mediterranean.[9] The other is Dalyan, in Turkey, a historic turtle breeding ground on whose behalf a small British tour firm called 'Turkish Delight' enlisted the support of leading British conservationist David Bellamy in their fight to prevent the construction of two large hotels. Bright lights from such developments would risk luring newly-hatched baby turtles inland instead of making the journey to the water's edge on which their survival depends. Turkey's fear of bad publicity just at a time when the country was desperately wanting to enter the EEC caused the government to agree on a compromise and a smaller hotel with shielded lights will now be built.

The plight of seals and turtles has the highest media profile but it is only the tip of the environmental iceberg. Undoubtedly the real problem is destruction of natural habitats, whether forest, river or sea. Local reaction to conservation efforts have not always been constructive. Volunteer wardens on Zakynthos have constantly been hassled and even assaulted by local developers, worried in case tourist development potential is harmed.

THE ALPS

The pristine slopes, fresh air, sunny skies and healthy image of the skiing holiday, sold as the perfect panacea for urban stress and pollution, has a strong appeal, particularly for the city-dwelling Briton who sees little clean snow and has to work hard to achieve the outdoor lifestyle so fashionable in the 1990s. There are 50 million people a year, including 750,000 British, who now travel to the Alps for an annual skiing holiday, and it's a figure that's increasing at a rate of 5 per cent annually.

Skiing in spring is just the boost you need to revive your spirits after the long northern winter, and it's now within the financial reach of most people. It looks fun, it gets you fit, and it's a great holiday for all the family. So, why in 1988 did the internationally famous climber Reinhold Messner suspend himself from a cable car wired on Mont Blanc to protest against further alpine tourist development? Why has it been necessary for pressure groups such as Mountain Wilderness and Alp Action to lobby uninterested governments and tour operators to impose controls as a matter of urgency? Surely skiing creates considerable employment, generates wealth for the host countries, and is a long-term money-spinner?

As with all arguments related to the impact of tourism, yes, all these things are true. But there's a down side to it too, and the hard fact is that skiing holidays with all their associated infrastructure have been allowed to develop in an uncontrolled way for far too long. It's the scale that's now the problem, the sheer numbers involved. Skiing is now so affordable that people are despoiling the very things they set out to enjoy – the classic twenty-first century tourist trap, where tourism destroys tourism, is becoming a reality.

The to-date unrestricted expansion of downhill skiing that has resulted in 40,000 ski-runs and 14,000 lifts has turned much of the once-idyllic landscape of the Alps into an eroding, concrete-covered aesthetic and ecological disaster area. Steel pylons, revolving restaurants, overhead cables, lifts, tows, noisy snow machines, vast hotel complexes – is this still the stuff of Heidi? Yet the very people who baulk at the thought of going on a beach holiday and staying in a purpose-built concrete block will happily book into a newly-built hotel on the slopes. Why? Because skiing is traditionally more upmarket. And with all that snow, and all the little touches put on so thoughtfully by the tourist boards and tour operators (horse-drawn carriages; *gluhwein* nights – not at all similar to the equivalent sangria night in summer, of course), it's still a pretty postcard scene. Come in the summer though, and the photos aren't quite so attractive. Put simply, the clean, white snow, like a clever cosmetic concealer, hides the rash of unsightly, inappropriate development that one can see in the summer months, when the snow is no longer there as a useful mask.

The Alps, with their fragile montane ecosystem, are the most threatened mountain system in the world, according to the World Conservation Union, based, appropriately enough, in Geneva. The problems associated with the mass skiing holiday industry are both direct and indirect. Skiing can cause environmental damage, such as the erosion and ripping apart of fragile vegetation when sharpened skis cut into thin snow, or when tree branches are broken off to make impromptu slalom runs – a common practice among experienced alpine skiers. All this happens in an environment where the growing season is invariably short, and the capacity of the vegetation to recover is limited. The use of artificial snow-making machines extends both the season and the problems, while forest clearance to make way for the ski-runs and lifts results in increasingly severe landslides – often (as in Morignone in the Italian Alps in 1987) with tragic results.

Indirectly, the problems are those which stem from the associated infrastructure of any type of mass tourism: construction of inappropriate hotel development and sports facilities; pressure on the infrastructure to cope with mass transportation and waste problems; the cultural impact that tourism brings to a region dominated by foreigners. A total of 250 million holiday days are spent in confined Alpine areas. This situation, unless very carefully managed (and there have been no effective strategy plans yet developed), would cause a problem wherever it was located. Hence the tragic landslides, eroding mountains and footpaths, vast quantities of waste with no proper infrastructure for long-term disposal; acid rain at unprecedented levels due to carbon monoxide pollution from the cars which throng to the slopes. The enormous increase in motor traffic and the huge motorway building programme to process the 40 million tourists per year has caused considerable problems. Traffic in 1955 on the Brenner Pass stood at 1,500 cars a day; today it is 55,000 a day in peak season. The level of salt-spreading, necessary to thaw ice, has polluted groundwater and killed vegetation, (what

The Negative Impact of Tourism 43

vegetation there is left, that is). Fifty per cent of what's left of the mountain forests have been assessed as damaged, sick or dying, by the Swiss Forest Agency, who are now less able as a result to control avalanches, stop falling rocks and inhibit earth slides. In Lake Maggiore a few years ago swimming was banned because the water was so polluted. Fish populations have been drastically reduced, and touring skiers are no longer advised to melt ice for drinking water.

The level of quarrying 'necessary' for the construction of yet more hotels and restaurants is changing the very shape of the Alps, as well as causing considerable damage. Above many of the resorts in Austria and Switzerland (in France and Italy, they don't even bother) are ugly, expensive avalanche barriers, giving the protection that the forests would naturally give, if left to grow undisturbed.

Between Visp and Brig, the quarrying of gravel has reduced an alluvial fan to a lunar landscape. When roads are constructed above the tree line, as happened at Valpelline in the Valle d'Aosta in the Italian Alps, the problems become critical. Sheet erosion often follows the installation of ski lifts, not only causing a serious new avalanche risk never mentioned in any brochure, but also reducing the amount of grazing land available to Alpine farmers, as has happened in Sorebois near Zinal. Many Alpine farmers are now abandoning the land in favour of a life without the threat of avalanche.

The newfound rich kid's pleasure – heliskiing – has a large environmental question-mark over its head. Skiing off-piste, way above the tree line, where the environment is at its most vulnerable, may be fashionable, suave and adventurous, but it's also destructive.

Where the waste from the tourist-industry side of skiing goes is yet another can of worms. Despite 'no dumping' signs, in the upper Rhone, above Fiesch, so much solid waste has been dumped in the last five years that the shape of the river bed has changed; vegetation has died and little but weeds remain. In Italy mountain restaurants dump their waste down the slopes; on the Marmolada the summer skiing pistes were actually being filled in with dumped polythene!

Recent mild winters, light snowfalls, and the wish for a longer skiing season have precipitated desperate measures: artificial snow is now regularly pumped onto slopes using massive machines. Apart from using vast quantities of water, thus risking draining local water resources and upsetting hydro-electric power generation, this also keeps the mountains under cover for unnatural lengths of time, and upsets the lifestyle of hibernating creatures. The current row in Verbier stems from World Wide Fund for Nature objections to the proposed construction of two artificial snow systems. All these things in themselves might not seem catastrophic, but the cumulative effect is just as worrying as what has happened to the Spanish costas in the last twenty years. Of course, it's not just a European problem. In Scotland, planned skiing developments in Lurcher's Gully in the Cairngorms have been scrapped after heavy pressure from a strong environmental lobby fearing further damage in one of Britain's last remaining wilderness areas.

The Himalayas, as any experienced trekker will tell you, 'burned out' years ago. The problem there is one of desertification of high-altitude valleys and deforestation as locals cut down trees to provide fuel for cooking and heating to accommodate their visitors. Rare fauna and flora is put under pressure as vegetation decreases. In the same way, in the Alps, with development creeping to higher altitudes each year to take advantage of longer snow cover (and therefore a longer, more profitable season), the problem gets worse by the year.

There is also, of course, the cultural aspect. The influence of a populus made up of 50 per cent second home-owners who come for a few, brief weeks a year has to be considered in any full assessment of skiing's impact on an area. In Courchevel and l'Alpe d'Huez the cultural effect of mass tourism on the local Alpine people has been extremely marked. The whole peasant way of life has been lost, claim the older generation of these regions. Of course, in the year-round, purpose-built resorts of les Arcs and La Plagne one doesn't have the problem of cultural loss because there was no indigenous culture there in the first place. However, now these resorts have been established for the best part of one generation, even the residents whose livelihood is dependent on the tourists resent the fact that they cannot properly establish an identity which will survive the disproportionate resident/tourist ratio that now exists in these areas.

These issues have been raised off and on now for the last decade, but what's being done? Basically, the Swiss, Germans and Austrians are taking increasing measures to protect the environment; the French and Italians less so. The difference in planning controls is considerable; so, of course, more developers are being attracted to the areas with least controls.

The Swiss were the first to decree that all cars must pass exhaust emission tests; they have banned cars in Zermatt, Kuhboden, Riederalp, and they now strictly limit the level of pollutants in the atmosphere. They have also re-seeded some of the mountain 'deserts', and are cleaning up lakes, and re-vegetating quarried sites. It is virtually impossible for any further development to take place in Swiss skiing resorts. The federal government must pass any application for further ski lifts, cable cars, and chair lifts; snow guns are rare in Switzerland.

The Austrians, as revealed at their Interski 91 Congress, are undertaking a programme including:

* a clampdown on future development of ski villages
* staggering peak holiday periods with the co-operation of schools
* abandoning Saturday to Saturday bookings for a more flexible system
* using railways more than cars to reach resorts
* imposing severe restrictions on off-piste skiing
* making ski instructors teach respect for nature as a part of their courses.

The Austrians, then, seem to be trying hard to counter the worst of the negative effects of the skiing industry, with everything from harsh restrictive measures (which curry no favour with most operators), to catchy PR phrases, such as the Vienna-based Director of the Austrian National Tourist Office, Klaus Lucas' slogan, 'We have not inherited the Alps from our grandfathers. We have borrowed them from our grandchildren'.

What options are really open to the tourist then? A constructive suggestion would be to consider *lang-lauf* (cross-country skiing).

This can be just as exhilarating as downhill skiing if you choose the right location, and can be enjoyed in the forests without causing them environmental damage. However, if you are still intent on downhill skiing in the Alps, find out in advance what the carrying capacity of the resort you select is (i.e. the number of people an area can hold and cater for without environmental despoilation). The London offices of the Tourist Boards will tell you what last year's figures were for that resort. Then contact Alp Action or Mountain Wilderness and ask their opinions on whether that resort is near to exceeding its capacity. Also check up on your tour operator.

Try not to go when the pressure on the slopes is at its greatest. The French school holidays (15 February to 15 March) should be avoided at all costs – for one's sanity as much as anything else! Tourist Boards are trying to persuade people to visit outside peak periods by offering *semaines blanches* (low tariffs). Use these.

Don't sharpen your skis to the 'nth' degree – the effect on thin snow is to slice the vegetation underneath in two.

Don't break off branches of passing trees as slalom poles. The reasons are self-evident.

Avoid high altitude resorts which use snow cannons when the natural cover doesn't live up to expectations (the tourist's expectations, not nature's). Choose resorts which bravely went ahead with a complete ban on cars. Also consider Alp Action's prototype reserve, La Lauzière (near Celliers) which combines top-class skiing with conservation issues, and actively encourages respect for nature amongst its skiers.

Of course the Alps and the Mediterranean are situated mainly within the developed world, even if they are relatively undeveloped parts of that world. In the Third World, areas like the Himalayas, where the planning controls and infrastructure may not be in place, the capacity to cope with visitors may be even less. Such countries are often asked to make the leap from a predominantly rural-based economy into a tourist-oriented set-up without passing through all the stages in between. Recent tourist developments in Goa and other southern Indian beach resorts have brought the package-holiday market to the area. Tourists more used to resorts such as Benidorm and Majorca are moving on to the new 'exotic' locations, pushing up the price of food, using up the limited and valuable water supplies, forcing fishermen off the beaches and generally disrupting the traditional pattern of life and local culture. Locals have responded to this by demonstrating and boycotting tourist hotels and beaches. Soon they say there will be no beaches left for Goans to walk on.

The Relative Impact of Different Types of Holidays

Obviously the type of holiday you choose to take will have a different impact on the country you visit (Chapters 5 and 6 explore this in more detail). For example, do two people on a standard two-week package to Benidorm or Majorca have a more negative tourist impact than a couple on a wildlife safari in Kenya or on a cruise in the Caribbean? The answers are not straightforward and vary depending on the circumstances. This is why we don't do green star ratings of activities. You can't make blanket statements like 'all cycling or walking holidays are green'. Mountain biking up an eroding hill, or walking up a crumbling footpath is *not* green! It depends how, when and where the activity is taking place.

A typical package has the merit of concentrating tourists in resorts built to cope with their demands, thus limiting their impact on other areas less able to cope. Journeys to sensitive environments in search of wilderness, wildlife, adventure or cultural sights can have a serious negative effect simply because such environments are often fragile, delicate and sensitive to any imbalances or pressure from visitors. Erosion caused by large numbers of people walking on mountain footpaths in the US National Parks, visitors queuing to see lions in East African parks, tourists seeing the sights in Venice, all of these types of tourist activity which might be considered 'better' holidays, in the sense of being more active or educational, may in fact have a greater negative impact than a package holiday in a purpose-built – already spoilt – resort. It's a question of the capacity of any environment, place or culture to cope with visitors and their activities: there is obviously a limit to this and once this is exceeded then damage will occur.

Historically there has been little recognition of the economic importance of tourism. Planning, organisation and control of tourism by governments has generally been on an ad hoc basis, geared towards increasing tourism revenues but without putting such development into longer-term strategies that ensure a wider perspective. More attention is paid to declining traditional industries than to the expanding services sector. Until tourism and related activities are taken more seriously, the true weighting of the relative values of the various benefits against the costs will be difficult to achieve.

There has been a growing realisation that tourism should not be left to market forces alone, but rather should be more subject to environmental planning and management. As the United Nations Environment Programme put it in 1986:

> Such a growing and enormous trade in tourism was bound to have (and has had) far-reaching environmental and social consequences. Generally in the past, tourism was not particularly well-planned or managed, much of it having been allowed to develop *ad hoc*, with facilities, including hotels, restaurants, water supply and electricity having been hastily erected or brought in as demand arose. Too often the inflow of tourists, coming *en masse* during a short holiday season, would overload the system, leading to water shortages, sewage problems, electricity cuts and overcrowding. Undoubtedly the seasonal nature of tourism exacerbated the infrastructure problem because of the high costs of expanding the supply side to cater for what was little more than a few months of the year. But with the development of competition and choice, the tourist has tended to become more discerning, requiring better value for what in effect is an extremely important portion of his or her working life. Less and less therefore will the tourist accept overcrowded beaches, shoddy hotels, inadequate facilities, noise and pollution. Moreover, as those dependent on the tourist trade have

come to realize, the packing in of more and more people into a resort area will ultimately lead to its losing its attraction.

The concept therefore arose that tourism should be subject to environmental planning and management, taking into account the well-being of the local population which too often has had to accept a large influx of tourists without having had any voice in such development.[10]

The recurring theme, as you will by now be aware, is that, in general, the negative impacts of tourism are closely related to the *numbers of people involved*. Quite simply it is the scale of tourist operations, or mass tourism, that has highlighted and exacerbated the problems discussed in the previous pages. More people mean more investment and more development with a consequent need to attract more people to earn a return on the original investment. And so the cycle goes on.

Tourist development generally, though not always, has been on 'greenfield' sites, i.e. previously virgin land. Consequently there is greater potential for direct conflict with the environment, the local landscape and the peoples' way of life.

Chapter 4

Tourism, the Provider?

'Travel has become one of the great forces for peace and understanding in our time. As people move throughout the world and learn to know each other's customs and to appreciate the qualities of individuals of each nation, we are building a level of international understanding which can sharply improve the atmosphere for world peace.' These fine words spoken by the late US President, John F Kennedy in his speech at the inauguration of the US Travel Service in 1961 sum up much of the optimism expressed in the 1950s and 1960s when tourism started to expand as a major industry. We have seen how many of these fine hopes have turned into enormous problems as the scale of change has despoiled both natural and man-made systems. That does not mean to say that tourists and tourism-related developments are all bad, but how much of the original vision has come true?

Tourism is a mixed blessing: it has great potential to be a genuinely positive force or it can result in the destruction of the environment, and the degradation of culture and the social order. The reason why tourism can be so destructive is not that there is anything wrong with the activity *per se* but lies rather in the way it is developed and managed.

There are undoubted benefits to be gained from tourism. It is worth looking at these to see what lessons they can teach us in terms of 'good tourism' and to see if the benefits actually outweigh the costs.

(A) ECONOMIC BENEFITS

Tourism as a massive boom industry obviously has an enormous impact worldwide. As we have already said, around 80 per cent of tourist activity takes place in countries within the developed economies of the world, countries where tourism forms only a part of a relatively strong and diverse economy. Indeed the perceived economic benefits are the major reasons why many developments are embarked upon. Investors – private or Government or a mixture of both – have put their money into projects with a view to making either a return on any development itself (generally the private investor's aim) or in stimulating the local economy through spending by tourists attracted to the area (generally the aim of governments). Such spending on items like hotels and accommodation, food and drink, entertainments and shopping for artefacts and gifts, filters down through the rest of the economy, encouraging all kinds of support services and industries. From a purely economic point of view there have been numerous successes.

In the 1950s the rural-based economies of Spain, Greece and Turkey were backward in comparison with northern European countries. In Spain tourism grew so rapidly that, by the mid 1980s it was bringing in around US$13 billion annually in foreign exchange, paving the way for a new lease of life and confidence in a nation traditionally amongst the most under-developed and reactionary in Europe. The rapid economic development has been made possible to a large extent by tourist income (21 per cent of total foreign earnings) which helps balance an otherwise persistent trade deficit.

Turkey experienced rapid economic growth in the 1980s to the extent that foreign exchange earnings tripled during the decade, helped largely by the tourism boom. This achieved the objectives of Turkey's military government which, in 1980, developed a new tourist strategy. The aim was to use tourism to bring in foreign currency, encourage foreign investment and create jobs. Cultural and environmental considerations were very much subordinated to these aims. In Greece the boom occurred mainly in the 1960s and 1970s and although still one of the most popular destinations it has lost out in the 1980s due to competition from Turkey and saturation of popular resorts.

For many Third World countries tourism has been embraced as a potential source of scarce foreign exchange needed to pay for imported materials and goods. As many developing countries suffered as a result of declining trade and crippling fuel bills in the 1970s, so they increasingly turned to tourism.

Closely linked to development is, of course, the question of jobs. The service industries – accommodation, transport, food and drink, entertainment – are generally labour intensive. Something like 100 million people worldwide rely, directly and indirectly, on tourism for their employment. Undoubtedly many of these jobs involve menial work; conditions are hard, hours irregular, employment is seasonal and the wages are low. But in countries where unemployment is high, where the labour pool is greater than the work available and where large-scale labour intensive industries do not exist or are mechanising or closing down, tourism is a positive force. In addition these jobs are generally provided in rural coastal areas where, traditionally, young people have left and drifted to the cities to find work. They can help prevent decline and maintain local communities in a vibrant and healthy state.

Tourism is used to manipulate the employment distribution all over Europe, most notably in Austria, Switzerland and in the Federal Republic of Germany, where development is actively encouraged in rural areas to alleviate a depressed agricultural economy. Diversification into tourism enables a local economy to become less dependent on a single producer and as an area becomes more attractive to employers, migration to the cities can be halted and the local community stabilised.[1]

The development of an area for tourism not only involves the building of new hotels and leisure facilities but also creates a need to improve existing public services and local amenities. In other words a government or local authority must invest in the infrastructure of the area to be developed if it is to meet the demands of its new tourist capacity. This investment is something which was neglected in the early days of expansion for mass tourism and many of the environmental problems are the result of poor planning in such areas as sewage and public transport. Of course improving the infrastructure of an area will benefit the local community for whom such investment is only made possible by the potential economic returns of tourism. The Turks have upgraded their airports as a result of the tourist boom and similarly the provision of an international airport in many Caribbean Islands has improved communications with the outside world.

While tourist developments in the Mediterranean have undoubtedly contributed to pollution there, far worse damage has been caused by agriculture and industry. Indeed, it might be argued that a result of the concern over the effect of agricultural and industrial pollution on tourism is a growing awareness of 'green' issues and those countries that resolutely ignore their environmental problems are finding that it is the tourists, not the problems, that go away. A national opinion poll conducted in 1986 found that although only 8 per cent of people felt environmental issues were their most pressing concern, when asked to comment on specific issues 79 per cent were worried about dirty beaches and bathing water. By 1989 the percentage of people most concerned about the environment had risen to 40 per cent and a Friends of the Earth study revealed that fears for the environment were second only to those about nuclear power.[2] The recent typhoid outbreak in the popular Spanish resort of Salou had a dramatic impact on its valuable tourist business. Bookings were down a staggering 70 per cent the following year! It is statistics such as these that cause existing facilities to be upgraded and extended and new resorts to realise the importance of adequate investment in the infrastructure.

Schemes set up to monitor and reduce specific areas of pollution, such as the Blue Flag Award run by the Foundation for Environmental Education in Europe, receive government backing when the tourist industry is at stake. Anything which has the potential to reduce such a valuable source of foreign exchange receives priority treatment. Benidorm, one of the first places to bear the brunt of mass tourism, has responded and the Spanish government has taken steps to cope with the demand. Benidorm currently boasts three Blue Flag awards for clean beaches and has been named as one of the eight cleanest beaches in the world.[3]

The Gallup poll, commissioned by the Tour Operators Study Group in conjunction with the tourist offices of Greece, Italy, Portugal, Spain and Tunisia, showed 86 per cent of people were concerned with the environmental standards of their holiday resorts.

And 74 per cent of the 1,500 questioned said this concern would influence their decision to visit that type of resort again. The survey showed that clean beaches and good hygiene standards were the two most important factors for the enjoyment of a holiday.

Some pioneering Third World tourist projects have helped to combat existing problems such as an inadequate water supply. One such project in Lower Casamance, Senegal, aimed to create locally-managed tourist villages, planned and run by cooperatives. To set the project in operation wells had to be drilled because the existing water supplies were close to drying up. A relatively small amount of investment made this possible and the resulting new wells were warmly welcomed by the villagers who would otherwise have had serious problems when the old wells dried up.[4] The creation of a new tourist resort at Ixtapa on Mexico's Pacific coast has been combined with an extensive programme of environmental improvement of the neighbouring town. This programme includes the installation of running water, a sewage system, tree-planting and coastal protection, paved roads and electricity supplies and the implementation of mosquito control, all of which will make a marked difference to the standard of living of local residents. Similar projects are under way in Tunisia, Senegal, Gambia, Bangladesh and Korea, where the environmental improvement programme also includes plans to provide irrigation to surrounding agricultural land.[5]

(B) IMPROVED HEALTH CARE AND SERVICES

Tourism development can give a much-needed impetus to tackle particular problems of health and disease which may otherwise prove limiting to any proposed developments. The programme at Ixtapa is an example.

(C) PROMOTION OF CULTURAL INTERCHANGE AND TRADITIONAL ARTS AND CRAFTS

Undoubtedly many of the arts and cultural traditions have benefited to a great extent from their promotion as tourist attractions.

Overseas visitors are drawn to Britain by such cultural attractions as the Edinburgh International Arts Festival, the wide variety of theatrical entertainment in London's West End and Britain's literary heritage. We in turn travel abroad to experience the Oberammergau Passion Play or Rio de Janeiro's Carnival. The tourist benefits from personal development, education and insight that such an experience can give, while for the host the tourist interest provides an impetus which sustains a tradition which might otherwise die out. Much of the flowering of arts, festivals, exhibitions and international competitions is due to the power they have in attracting tourists. During the Cold War years, tourist travel between East and West helped to promote understanding and cooperation by providing the opportunity to overcome media-inspired stereotypes and national prejudices. In Bali tourism has promoted a revival of some traditional arts and crafts: in 1973 there were 4,000 wood carvers practising their craft on the island and their products were not only sold to the tourists but also exported.[6]

On Nias, an Indonesian island off the west coast of Sumatra, a traditional local dance to intimidate and then welcome visitors to the island is once again performed by the men and women of the mountain villages. In Tunisia the crafts involved in jewellery manufacture and leatherwork have gained a new importance as have the traditional textile and glass crafts of the Maltese.

Of course, while encouraging the positive impact of tourism on local culture it is important to differentiate between authentic native cultural traditions and cheapened trivialised versions such as the 'tartan haggis' shows in many Scottish hotels. It is not uncommon to find traditional Sumatran marriage ceremonies held at eleven o'clock every Tuesday for the sole benefit of the tourists to Sumatra.

(D) PROTECTION OF HISTORIC AND CULTURAL MONUMENTS AND BUILDINGS

Tourism has been the stimulus behind the preservation and conservation of many of the great historic and cultural monuments and buildings throughout the world. In Britain and Europe great country houses, palaces, castles and chateaux have been successfully converted into hotels, museums, galleries and visitor centres. Many of these buildings were in a severe state of disrepair and in search of a new use. Imaginative restoration of such places as Ayudhya, the ruined former capital of Thailand, Anuradhapura, the ancient Sri Lankan jungle city and the Buddhist temple at Borobudur in Java have saved these cultural heritage sites of world importance from further deterioration. Spectacular examples of large-scale restoration work include attempts to prevent Venice from sinking and remedial work in Florence following the disastrous flooding of the River Arno. One great conservation success story is the rescue of the ancient Egyptian temples and buildings of Abu Simbel on the Nile where the whole complex was moved stone by stone to save it from rising water levels following the construction of the Aswan Dam. Bearing in mind the counter examples presented in Chapter 3, showing the negative impacts of tourism on cultural and environmental areas of interest, it is clear that we are dealing with a paradoxical beast: tourism can be both a saviour and a threat.

(E) CONSERVATION OF WILDLIFE AND HABITATS

While acknowledging that tourism is primarily an economic activity, there are outstanding examples of how economic benefit can be successfully married to the conservation of individual endangered species of wildlife and the protection of special or important wilderness habitats. In Kenya the safari business is now highly organised. Shrewdly, the Kenyans have realised just how valuable their wildlife is in terms of its potential to attract tourists and revenue. Fleets of buses visit the game parks and the wildlife areas to view the lions, zebra and wildebeest. So long as tourism continues to flourish and provide revenue so the threat from population pressure to cultivate the potentially rich agricultural grounds within the parks will be kept at bay. The realisation of the tourist potential of observing elephants in the wild is also serving to encourage protection of these magnificent animals. However, there is a danger that too many visitors might come to the Parks and some areas such as the Masai Mara National

Tourism, the Provider? 57

Reserve show signs of having reached their capacity and cannot continue to support such activities without seeing a detrimental effect on the wildlife itself. Deciding when enough is enough is the crucial question.

One of the most dramatic success stories in recent years involves the rare mountain gorillas of Central Africa, on whose behalf a campaign was fought against poachers by scientist Diane Fossey which eventually cost her her life. Her story, immortalised in the film *Gorillas in the Mist*, first brought the plight of the rapidly disappearing primates to the world's attention and resulted in the setting up of the Mountain Forest Project in Rwanda. This project adopted the strategy of encouraging tourism in order to bring desperately needed money into a poor country to fund their work. Gorilla tourism is now the second most important source of foreign currency in Rwanda. Tourism is strictly controlled and visiting groups are limited in number to prevent disruption to the gorillas' natural environment. The project's administrators are now looking to use the programme to protect species in other countries such as Uganda.[7]

In the Caribbean Cayman Islands a government-owned turtle farm has been established to breed green sea turtles and release them back into their natural marine environment. In order to fund the project a small proportion of the farm's massive output of turtles are exported for tortoiseshell. This compromise situation allows tourists to observe and understand these animals within the farm and also to protect their natural environment. The whole coastline has been designated a marine park and fishing and interference with the coral reef are banned.

Outstanding examples of conservation campaigns from Asia and the Pacific include Ste. Anne National Marine Park in the Seychelles, where the park is used by both residents and tourists for swimming, sailing, snorkelling, diving, and glass-bottom boat excursions; Chitwan National Park in Nepal, where tourism developments have been kept within rigorous limits and tourism has been a major justification for saving the endangered Great Indian Rhinoceros; and Tai Island in Fiji, where as a result of protection, subsistence fish catches have increased, tourist activity has expanded, and the holders of traditional fishing rights are involved in resort management and boat hire.[8]

The boom in recent years of holidays with a natural history interest such as bird-watching tours and study tours to National Parks, which deepen understanding of the indigenous flora and fauna, is a measure of the interest and concern many people now feel for the environment. Provided such tourism is undertaken in a sensitive way it can be a powerful communicator of the need to protect our environment in general and the special wild places that still exist, in particular.

The influence exerted by the tourist and leisure lobby can be especially influential in preventing inappropriate development in many popular areas. In the Lake District building standards have been improved and major road-building projects prevented as a result of pressure from the Friends of the Lake District group who are determined to maintain the character of this beautiful area. The threat to tourism can be an ally in such circumstances as it lends a commercial counter-argument to the debate between what was hitherto economic development and progress versus aesthetic and cultural values.

In India – a country where the population pressures to expand into wilderness areas are enormous – tourism is being seen as the key to conserving wildlife. Conservationists believe that controlled visits by tourists genuinely interested in wildlife and its long-term survival (ecotourism) are the best chance of saving its habitat. Income generated in this way can be more valuable than clearing the land for agriculture. The Indian branch of the World Wide Fund for Nature considers that 'the only lobby for conservation in India is the tourism industry'.[9] In the Kaziranga National Park in Assam, the loss of twenty-five one-horned rhinos through poaching was blamed partly on the absence of tourism in the area, as the resources available to protect the rhinos were obviously inadequate. The report concluded that tourists should be encouraged, if only to help the authorities find out what is happening in the park.

Unfortunately tourism development is seen more and more as a major threat in itself. In the debate on whether a development is appropriate or not there is rarely a clear cut right or wrong answer.

Take the case of the green sea turtle in the Cayman Islands. Many would argue that such exploitation of a wild animal for profit and human fashion is distasteful, morally wrong and, by creating a market, likely to encourage indiscriminate poaching of wild populations. On the other hand careful management and direct intervention may be necessary to help conserve the animals and the compromise of farming some and releasing others to the wild will help maintain the species. Such dilemmas are common and represent the difficulty of finding realistic solutions which are practical and sustainable and also likely to satisfy the majority of interest.

There is a decided element of hypocrisy when rich countries voice conservationist concerns about areas they do not live in and of which they know little of. Those who live in Third World countries or less developed parts of the developed world such as the Scottish Highlands will often have a quite different perspective on what is appropriate for their own environment and lifestyle. Tourists and tourism may represent the opportunity to escape from the grinding poverty or hard work often necessary to earn a living. This is how many locals see it. Clashes are increasingly common. For example, in Lurcher's Gully in the Cairngorms of Scotland proposed skiing developments generally enjoy the support of locals and government officials, but they have been vehemently opposed by conservationists and non-governmental national bodies. In Cyprus measures to protect the Akamas peninsula from careless development by restricting the tourist capacity and limiting the scale of development are meeting with surprising opposition from local villagers who have witnessed the fortunes made by their southern neighbours in selling land for tourist development and feel aggrieved that they are being prevented from doing the same. Having become accustomed to the stereotyped sun/sea/sand packages that have always attracted tourists by the thousands, local people everywhere need convincing that green tourism projects will actually be successful and prove financially, environmentally and culturally profitable.[10]

In this chapter we have cited numerous examples of the positive side of tourism. It would be hard to deny its importance to the economy of many countries, as it has broadly encouraged development in many parts of the world which previously had no external international economic base. It is also a means of stimulating people to visit and learn about other countries and environments and, by being a tourist, help give a value to those things – cultural diversity, natural environment – which make the world the place it is. Despite, or maybe because of, the world trend for global homogenisation, cultural diversity is thriving, revived in many areas by the desire of tourists for 'authenticity' and participation in or observation of local events and lifestyles in a non-artificial manner.

The trick, of course, is to ensure that tourism, local culture and the environment are in balance. There are no easy answers to this. Chapter 3 illustrated the problems when these factors are not in equilibrium. Tourism can, however, be a catalyst in providing solutions to environmental problems by offering alternatives to traditional forms of land use and economic exploitation. The examples of ecotourism, such as wildlife safaris and whale study cruises, are, when properly undertaken, viable ways of valuing and protecting the very resource the tourist comes to see. The key lies in applying the principles of sustainable tourism development, which guard against the enemies of a healthy balance between developers with entirely selfish aims, whose time-span is short term and sole motive financial, and those radical environmentalists who are unwilling to brook any compromise and who, by their own intransigence, encourage intransigence in others. The principles are as applicable to the Mediterranean resorts as they are to the game parks of East Africa.

Chapter 5

The Way Forward

Tourism and Society

In the last couple of years, the whole question of the impact of tourism and its role in influencing society and the environment has been given a more public airing. For example, in 1991 the UK government published the findings and recommendations of its Task Forces looking at the impact of tourism on the environment in the UK. In 1990, ninety-two experts representing twenty-one countries met in Washington for the First International Assembly on Global Tourism Policy. The overwhelming message from this body of experts was that tourism's environmental impact was the key issue of the 1990s; the tourist industry awarded its first 'Tourism for Tomorrow' awards. The President of American Express Travel Related Services stated in 1991 that the environment is the one issue which will, if left unexamined, pose the greatest threat to the long-term growth of tourism. Many tour operators have also rushed to demonstrate their 'green credentials'.

So the problems appear to have been recognised, but how seriously are they being taken, and what action is taking place to make realistic progress in reconciling the conflict?

It is always easier to analyse and diagnose problems than to suggest possible solutions. It has become rather fashionable of late to criticise tourists and tourist development in terms of their detrimental effect. There has always been a tendency to view tourists smugly as some sort of inferior beings, who require being led by the hand and cushioning from alien food and culture – intent only on the selfish pleasures of 'having a good time'. After all, tourists are always someone else and never us. The development of tourism on its current scale presents problems and opportunities: many people have diagnosed the problem and paint a gloomy picture of tourist hordes destroying the last undiscovered places. There are undoubtedly serious problems but can they not be tackled?

How practical and realistic are the potential solutions? There has, at last, been a realisation by sociologists, tourism analysts, developers, entrepreneurs, operators and tourists themselves that there are major problems: the recent drop in bookings to traditional locations such as Spain and the change in attitudes of tourists and the type of holiday being pursued are evidence of that. The fact that there is a debate on tourism and its effects is at least a large step forward. It is now at the forefront of the agenda for the industry as a whole. Cynics might argue that the industry – travel agents, tour operators, hoteliers, developers etc. – are only superficially responding to what they see as a new niche in the market. However it is a start and the industry will ignore current trends in public awareness at their peril. The environment is uppermost in many people's minds and will undoubtedly influence the way people choose to spend their holidays.

This chapter explores some potential solutions to the issues and dilemmas currently facing tourism. These will be mainly from the perspective of the industry itself, the external influence on the tourist. Chapter 6 will look at the way the tourist can influence the industry and help to contribute to a 'better' form of tourism by their behaviour – the internal influence. Realistically it is only by these two forces coming together – from the top down and the bottom up – that any tangible change can occur.

The future of tourism cannot be separated from the society in which we live. It's very tempting to see the way in which we spend our leisure time as completely unconnected with the rest of our lives because one of the main reasons for travelling away from home is to put the cares of everyday life aside for a while, to indulge in a temporary world of pleasure and relaxation that has nothing to do with our normal routines. The vital connection between the conditions of our daily life and how we choose to spend our leisure time is usually missed or ignored. An increasing world population with an ever higher standard of living, combined with the advancements of science and technology that have brought the world to our doorstep, mean an increasing number of people able to travel are doing so. The greater materialism of the twentieth century has produced a disposable consumer culture, with short concentration spans and low boredom thresholds. We are becoming used to the unusual and this is compelling us to search for things that will still surprise and refresh us. As creatures with natural animal instincts we find the modern lifestyle suffocating if it is not relieved by a short annual burst of spontaneity and authenticity. Many travel further and further in search of this, attempting to get 'back to nature' and find somewhere 'unspoilt'.

Our society in the developed world is also increasingly geared towards free-market economics with a resultant focus upon self-interest that may be useful in the business world and market place but which encourages an exploitative and acquisitive approach to the environment. The tourist market should be in essence a selfless one, after all tour operators and agents are concerned to provide their customers with what they want, so why should this concern for the desires of others not extend to the host communities? In the end it boils down to profit. A tour operator depends on both the host community and the tourist for his business and neither can operate without the other. However, displeasure felt by customers has a far greater adverse effect on profits than that suffered, often in silence, by the hosts.

It is these factors, then, which influence and determine our travel habits and any attempt to develop a more responsible attitude towards tourism must take these into account. Tourism does not exist in isolation from all other aspects of life, rather it is closely linked with everyday economic and social concerns and it is only by addressing these issues that we can begin to tackle the roots of the tourism problem.

Sustainable Development

Current trends towards environmental consciousness and the rising popularity of the 'green' movement in all its forms show that many in the world are becoming aware of the mistakes that have been made in the past and of the need both to correct these wherever possible and to prevent them being repeated in the future. However, this awareness alone is not enough. Definite action needs to be taken immediately and on a massive scale to deal with the urgent problems. We must learn how to manage the mass tourism phenomenon rather than simply try in vain to accommodate it.

Most importantly countries must begin to apply the principles of sustainable, long-term development to their existing or proposed tourism projects. The Word Tourism Organisation's 1989 Hague Declaration outlined these principles:

> Countries should determine their national priorities
> and tourism's role in the 'hierarchy' of these priorities
> as well as the optimum tourism strategy, within these
> priorities. This strategy should define, among others,
> the balance to be sought between international and
> domestic tourism and take into account the carrying
> capacity of destinations . . . Within the overall national
> tourism strategy, priority attention should be given
> to selective and controlled development of tourist
> infrastructure, facilities, demand, and overall tourist
> capacity, in order to protect the environment and local
> population, so as to avoid any negative impacts which
> unplanned tourism might produce. In tourism planning
> and area development it is essential for States to strike
> a harmonious balance between economic and ecological
> considerations.

The WTO has been pushing environmental concerns since the 1960s and has done much associated research and development work, but little of their work has been implemented. This is mainly because important industrialised countries such as the UK, USA and Australia, representing the main generators and investors in tourism, are not members.

So it's all been said before, but how many people working in the tourist industry could say they have read the Hague Declaration and how many have even heard of the WTO? Their guidelines, however, emphasise the particular importance of the carrying capacity of a destination. There is a need to define and implement policies which place limits on the maximum number of visitors and any development which is undertaken to house and amuse them and there has to be at least an equal priority given to the well-being of the local population.

In the impulsive rush of development for economic profit so often the environmental impacts on a village, town, region or island are ignored because to heed them would be to place a limit on development and consequently on profits. Little value is given to such factors as clean beaches, fresh air, local cultural values, wildlife and other intangibles. Other factors such as the size of the local resident population, the availability of resources such as water and electricity and the resilience of the environment should be taken into account before development takes place and, ideally, a limit placed on tourist numbers that reflects this carrying capacity. Until now, where such limits have been imposed at the outset, subsequent popularity and demand have led to a gradual raising of these quotas until the whole idea of a capacity total is a farce. For example, an original limit of 12,000 annual visitors was placed on the Galapagos Islands, home to many unique plant and animal species, in order to protect this fragile ecosystem. Very soon this limit was breached and a new limit of 25,000 imposed. In 1986 over 30,000 visitors set foot on the Islands, showing that once again restraints proved ineffective in the face of profiteering.[1] David Attenborough in his book *State of the Ark*, states his view that integrated, sustainable tourism is possible if simple, sensible precautions with regard to carrying capacity are taken at the outset:

> Tourism should be viewed as an ecological phenomenon.
> Tourists are like a new species invading an island.
> Because they are flexible, they are inclined to overrun
> the whole place, but if they are to persist on the island
> as they found it, they must settle down in a niche and
> behave themselves. The resident 'species' – the local
> people and the wildlife – may have to make some
> adjustments to accommodate the new neighbours, but
> eventually they should be able to use them to their
> mutual benefit.[2]

Of course what is really needed is the political will to take sensible precautions or even in some cases oppose inappropriate development. Many countries have appropriate safeguards in law to protect their environment and control development. Unfortunately the will to impose such control, monitor projects and act against transgressors is simply lacking. Every year in the Mediterranean many forest fires 'spontaneously' appear in protected areas which are then subject to development proposals to 'rehabilitate' the area. In Spain, home of some of the worst 'boom town' development, there has at last been a recognition of the need to rethink the policy of quantity rather than quality in terms of holidaymakers.[3]

Potential Solutions

Are there solutions to the problems of mass tourism and its associated developments? How realistic are they and what are their chances of success? The answers to these questions are not easy to find but they have to be sought. Indeed many individuals, organisations, companies and governments do recognise the need for change.

It doesn't take much foresight to recognise that ultimately the world will run out of new places. One solution of course is to do nothing and let the market place sort out the problems. In essence areas which allow deterioration of their environment and inappropriate development will be penalised by the consumer who will recognise the reduction in quality and take his custom elsewhere. Thus the industry will be forced to respond and make changes. In practice, in many countries the market-place has been the 'method' by which the tourist industry has been 'regulated', yet paradoxically, despite its economic significance, it has lacked the political clout to influence national policy due to its fragmented nature and ignorance of its true role and importance.

Problems over sub-standard accommodation, dirty beaches and seas, concrete vistas, overcrowding, disease outbreaks and airport delays have all had an adverse effect on the industry, forcing it to clean up its act or risk total collapse. This reactive mode can be appropriate in some instances. After all, central planning by governments to 'create and control' tourism has sometimes had the most disastrous effects as witnessed in several developed and Third World countries. Nevertheless there is a delay in the way that the market place acts and this delay invariably results in irreversible damage: can the Mediterranean ever be anything other than grand-scale tourist resorts? That does not mean, however, that the market place is not a useful tool in controlling tourism. The industry can, by flexing its economic muscle, act as a positive force for change by insisting that investment and development will not take place without appropriate measures or infrastructure and environmental safeguards to protect the attractions which inevitably draw the tourists in the first place. Disney World in Florida is a good example of the power of a large corporation to influence such matters.

At the other extreme is the proposition that all tourism is bad and that the only solution is to severely restrict all tourist development and its related activities – transport, construction and dislocation of people. In terms of solving many of the problems this is certainly attractive – no tourists, no tourism problem. But realistically and politically is it an acceptable solution? Prohibition rarely succeeds. Too many people are employed in an industry which generates too much revenue for it simply to be stopped. Anyway, many enjoy travelling and look on it as a basic human right, which indeed it is. However it does again raise the point of why be a tourist and why travel? A less glamorous image and a more realistic appraisal of the pros and cons might at least persuade some people not to take a holiday or travel to a far-off destination. They may be much happier doing something else.

Although it may not be practical to restrict all tourism, the management of numbers is critical to the whole concept of sustainable tourism. In some countries and places tourists are simply not encouraged. The small Himalayan state of Bhutan, sandwiched between two giant neighbours China and India, only allowed foreigners access for the first time in 1974 after 300 years of splendid isolation. Yet only fourteen years later the Bhutanese government imposed drastic reductions on the number of tourists allowed to enter the country – 2,400 a year – and recalled all but one of its overseas tourist agents. The decision was taken as a result of a report presented to the Bhutanese Parliament by a special committee for cultural affairs which analysed the effects of tourism on the local society and culture and concluded that, 'Tourism is harmful to the holy nature of our Buddhist monasteries, contributes to the desecration of the country's holy places and corrupts the population.' The World Bank estimates that Bhutan is one of the world's least developed countries and yet, as a country self-sufficient in rice, wheat and corn, and where most clothes are hand woven and sewn rather than bought, most of its 1.3 million inhabitants enjoy a higher standard of living than their Indian neighbours. Bhutan's King Jigme Singye Wangchuk is reputed to have told World Bank representatives that he took the decision to severely limit tourism in the interests of his country's 'GNH' – Gross National Happiness. The King has obviously decided that the carrying capacity of his Kingdom is very small indeed.[4]

Tourist and visitor management is an essential tool in any attempt to control people. It takes many subtle, and sometimes not so subtle, forms but is rapidly being implemented in many tourist attractions. At Lascaux in France a complete replica of the original cave with drawings has been reproduced close to the real one. Visitors get to appreciate the atmosphere of the cave by visiting the replica and thus not damaging the original. Venice is currently contemplating restricting numbers having had to close the city once already. The simplest way of course is to price the particular attraction out of the range of most people. However that is no guarantee of numbers or indeed the impact of tourism being lessened and it introduces an undesirable élitist element. Indeed, high price tourist activities such as heli-skiing in the Himalayas are increasingly popular and can often be extremely damaging. Nevertheless it is perfectly reasonable that tourists visiting sites and areas of natural interest should pay something towards conserving that site; if a tourism development will have an environmental cost then it is right that the developer pay towards protecting that environment. The principle of 'polluter pays' is well accepted in environmental economics and is gradually being included in legislation by many governments.

Unfortunately as far as tourists and tourism are concerned

many of the attractions of a resort or country – landscape, scenery, wildlife, sunshine, beaches, access to the sea and so on – have been provided 'free' or at no cost to the tourist or tourist developer. 'The elemental theory of supply and demand tells us if something is provided at zero price, more of it will be demanded than if there was a positive price. The danger is that the greater level of demand will be unrelated to the capacity of the relevant natural environments to meet the demand', as Professor Pearce so comprehensively put it in *Blueprint for a Green Economy*.[5] The result? – the overcrowding of many tourist resorts.

One good illustration of the role good management can play in reconciling such conflicts is the case of the Rotorua Geysers in New Zealand. The geysers have long been a prime tourist attraction, happily coupled with a source of cheap geothermal energy, providing heating for both hotels and private homes. From the late 1960s, however, there was a gradual decline in geothermal activity, particularly at Rotorua's most famous site, Whakarewarewa. As the geysers stopped spouting, the visitors stopped coming, the cheap energy source disappeared and the whole tourist industry was under threat. Central and local government agencies developed a management plan involving the closing of all geothermal wells within 1.5 kilometres of Whakarewarewa, which, after much controversy, was implemented. Though the tourism industry was forced to switch to other, more costly, energy sources, the conservation measures paid off. The geysers started spouting again, geothermal activity returned to normal, and the tourist attractions themselves regained their appeal. The judicious imposition of controls and additional costs on the industry which had grown up to benefit from these attractions was an example of long-term sustainable management which served to save the attractions themselves from irrevocable damage.

Undoubtedly we get our holiday 'too cheaply' in terms of environmental cost and the impact on the host population. On the other hand the market place 'instils' the discipline of competition which encourages countries to compete in terms of 'value for money'. This invites a short-term view and often the environment is the loser. Perhaps 'value for money' should be redefined as 'environmental value for money' so that all aspects are taken into consideration. Tourism will have to cost more in order to protect those very things that tourists value.

The Growth of Responsible or Sustainable Tourism

Response to this new kind of tourism has been quick to grow. Organisations like the recently formed Tourism Concern which focus on the impact of tourism on people living in holiday destinations, particularly those in the Third World, reflect the mounting interest. It has three aims: (a) to undertake and collate research on the impact of tourism (b) to share ideas and maintain a network for information exchange (c) to work for constructive tourism practice. The British travel industry itself is taking more than a passing interest. The 'Tourism for Tomorrow' awards have been introduced jointly by the Tour Operators' Study Group, the British Tourist Authority and the TV holiday programme *Wish You Were Here* under the chairmanship of well-known environmental campaigner Dr David Bellamy. These awards, started in 1991, are for tourism projects which show a combination of creativity, short- and long-term benefits to the local community and tourists and a balance between the needs of both parties.

Some have argued that 'responsible' tourism alone is not a real answer: it can be construed as a middle-class, educated and élitist cop-out for those who are caught up in the groundswell of green issues. It is small scale, slow and attempting steady, controlled, and thus sustainable, growth. Is it really the answer to the enormous problems of mass tourism dealing with the movement of enormous numbers of people? The answer is probably no, not in itself, but it can be part of an overall approach to tackling the problems. Such an approach must encompass a whole range of potential solutions each of which may be appropriate according to the situations and circumstances.

Is mass tourism inherently wrong? Well, large numbers of people visiting places, congesting the roads and airports, craving new experiences and using up more resources is a critical part of the problem but it has at least brought the opportunities of travel to very many people. By its very nature, large-scale tourism can only be coped with by modern marketing methods and the economies of scale, i.e. mass-production, high turnover, standardisation, high-capacity facilities and high-profile advertising and promotion. There are many places where mass tourism can be coped with perfectly adequately: Disneyland and Disney World, Alton Towers and Benidorm are basically honest and up-front in that they set out to be centres to cope with a large flow of visitors. They allow people to do exactly what many holidaymakers want to do – rest and relax and recuperate from their normal routines – and they fill a very large niche in the tourist market. The idea of tourist 'honeypots' readily identified and with the necessary infrastructure to cope with large numbers of people, is appealing. Sometimes these developments have been referred to rather sarcastically as tourism ghettoes – people choose to holiday in these resorts simply to relax and have fun, surrounded by others with the same aim and in an environment where everything is geared towards leisure. The ghetto idea openly admits that its sole aim is recreation and relaxation. Club Med's own advertising illustrates this:

> The idea behind Club Méditerranée is as old as the Fall of Man. It is the idea of a paradise, a Garden of Eden, in which people are free and unconstrained and everybody can be happy in his/her way, regardless of whether it is in a sunshine village in the Antilles or in a snow village in the most beautiful part of the Alps.[6]

By their nature these large leisure complexes are carefully planned and controlled. They are situated in isolation from their surroundings and the tourists rarely spill out to 'spoil' the local environment. They largely achieve their aim of giving people a good time, enabling them to return to 'normal' life refreshed and relaxed. Opponents of the tourist ghettoes, or 'honeypots' cite the absence of any genuine experience of the host country and the almost complete lack of economic benefit to the local community as arguments against the concept. It must be remembered, though, that a sufficient number of people just want to go and escape for 'sun, sea and sand'. It doesn't really matter where and if these people go to the holiday clubs and on the cruise ships, a lot of pressure is taken off the other more sensitive areas.

This is basically an honest formula for a holiday and may ultimately have the least negative consequences for the host country and its population. That does not mean that there could not be improvements from the point of view of the guest/host relationship, particularly in the Third World. That ideally will be the next phase.

The creation of custom-built ghettoes may be one way of dealing with mass tourism. The question of the development of areas with fragile natural environments, such as National Parks, mountain ranges, coastal vegetation, or tropical rain forest requires a different approach, however. Can places like the Lake District, the Alps, Westminster Abbey, Stonehenge or even whole cities like Bath or Venice be regarded as tourist honeypots, to be developed to cope solely with tourists? Surely there is a limit to the capacity of such places to cope with the huge influx of visitors and vehicles at peak times? The answer for such places may be one of dispersal, encouraging visitors to other less crowded and probably less well-known places or to different times of the year – in other words spreading the load over a wider season. Examples of moves in this direction already exist. We are encouraged to take winter breaks to European cities and given considerable off-season discounts. The WTO has intelligently proposed that all tourist-generating countries should stagger their industrial and school holidays, so that people have more choice of when they can travel. Many would actually prefer to travel at off-peak times in order to avoid such notorious holiday problems as traffic jams, long queues and airport delays. Such a step would avoid the situation that occurs in France where practically the entire population goes on holiday in August!

Care is obviously required to provide appropriate solutions. It is no good dispersing visitor numbers if they are merely spread to areas which cannot cope – all this will do is exacerbate the problem. Such concerns are currently being voiced in

Spain where, in response to the recent bad press and depressed booking of the traditional Costas, the Spanish government is pushing the cultural and physical attractions of inland Spain. This is perfectly understandable given the desire to raise the quality of tourist experience and the undoubted attractions of such wonderful places as Cordoba, Seville, Toledo, Picos de Europa and north-western Spain. But are these places geared up to cope with mass tourism? Is the necessary infrastructure in place? Will this destroy the very attraction of these places? There needs to be a rationalisation of what is suitable.

The Role of Governments

As previously discussed, there has been a tendency on the part of governments to discuss tourism as a 'candy-floss' industry and to consider that the jobs created were not 'real' jobs, compared, say, with those in farming or manufacturing. Of course, there was a commitment to attracting more visitors to spend money, but, because the benefits of tourism are not always obvious to the mass of the local community, when resources are scarce the political spotlight (and hence the allocation of funds) has fallen on other issues rather than those of tourism and visitor management.

Nonetheless, the role of governments in planning, developing, controlling and monitoring the tourist industry is vital. In the past, tourist development for economic benefit was often the main motivation for governments to get involved. They have often done this in conjunction with multinational investors in order to find the necessary capital. Too often such projects have proved dubiously beneficial in economic terms to the host country, never mind the environmental cost and social upheaval. At the planning stage full analysis of all the costs and benefits needs to be undertaken in some form of environmental impact assessment. Such assessment is not easy though the techniques are well-known and utilised throughout the developed world. Tourists in particular value such intangibles as clean beaches, pollution-free water, landscape, natural habitats and wildlife. It should be possible, therefore, to build such values into the costs and benefits of the assessment and assign a relative weighting to each. Developers might argue that the imposition of bureaucracy will impede an industry that has to be fast and dynamic in exploiting what is a very unpredictable market. However it is precisely because of such an approach that ill-conceived and inappropriate developments have caused problems. In the long run the industry itself will be the better for a long-term approach that attempts to encourage sustainable development and thus a sustainable market.

How might this be done? In many Western countries the local and national government infrastructure already exists to implement appropriate strategic planning initiatives encompassing these factors. What is required is for tourism to be recognised for its true contribution, with a resultant need to manage it properly to sustain the benefits and minimise the costs. In some areas, of course, tourist development in terms of hotel construction and encouraging more visitors will be entirely inappropriate, whilst other areas can happily cope. The private sector itself would welcome such coordination in terms of indicating the likely national or local view. Indicative tourism management would identify preferred, potential or sensitive areas – i.e. a form of zoning to indicate to developers, conservationists and the local population how tourism development might proceed. It might also indicate the form of tourist activity appropriate for a locality – e.g. whilst large numbers of tourists may be happily catered for in a custom-built resort, this might be entirely inappropriate for alpine or mountain areas where low levels of people and low-impact forms of tourism such as walking would be encouraged. In order to achieve such an approach, it will be necessary to have a database of the current state of the locality or country, identifying what is present and what its relative worth is, whether it be a rare and fragile ecosystem or a historic and interesting old town. Unfortunately, even in developed countries this rarely exists. The picture in Third World countries is often much worse. The EEC has at least recognised that the tourism sector has not received adequate financial support.

The example of the Algarve provides a demonstration of what

has been wrong in the piecemeal planning approach, and what might be a way forward. The Algarve covers an area of 5,000 square kilometres with a population of around 330,000 inhabitants unevenly distributed throughout the territory. It is the main tourist destination in Portugal and receives 1.5 million visitors during the holiday period. The haphazard nature of tourist development in the area has, however, caused fundamental problems in terms of the gap between the existing services provided – water supply, sanitation, transport – and the needs both of the local population and visitors. Tourist services have been unplanned, with little diversification. Investment is mainly in accommodation, 90 per cent of which is concentrated on the coastline. With the encouragement of the EEC, a Regional Development Programme was drawn up for the Algarve for the period 1986–90, with the aim of reconciling the various interests of regional government, local people and the tourist industry. This set clear objectives for tourist development in terms of infrastructure improvement, restoration of the cultural heritage of the region and the diversification of tourism away from just a few coastal honeypots.[7]

Planning and cooperation must extend across international boundaries. The problems of the Mediterranean will not be solved by just Spain cleaning up its act; all countries with a coastline on the Med, need to act. International cooperation is notoriously fickle but the Mediterranean Action Plan initiated in 1976 represents a remarkable inter-governmental consensus on the nature and degree of the principle problems that affect the region in terms of pollution prevention, environmental protection in general and management of wildlife and habitat conservation. The Action Plan has probably slowed the environmental decline of the Med but not halted it. What is needed now is to increase the awareness of ordinary people whose everyday lives are affected by the Med and by its overall health. Indeed this is the basic level in any plans to provide a more sustainable form of development and tourism – they must involve local people without whose cooperation all good thoughts and plans will be futile.

There are many organisations in place like the WTO who

have published recommended standards and guidelines for tourism development: it should always be 'Consistent with making maximum use of national and local resources but which avoid serious adverse social, cultural and environmental impacts.' Many Governments enshrine such well-meaning phrases in policy and legislation but lack either the inclination or the resources to implement intentions or monitor their usefulness. What is required is the will to turn such good intentions into actions. It would help if more countries (i.e. the UK) were members of the WTO. The process must start with the developed world because it is they who are both the chief cause and on the receiving end of most of the problems. They also have the resources to do something about it and aid poorer countries which require both guidance and example to believe that it is in their long-term interest.

Operators and Agents

There has been an enormous leap in interest in tourism and the environment in the last year or so. In the boom years of the 1980s, although many environmentalists pointed out some of the more questionable aspects of the tourist revolution, the industry itself generally felt able to ignore the warnings, safe in the midst of the holiday boom and buoyant profits. Now almost everyone has a view on the subject, and operators, hotels, airlines et al. rush to proclaim how 'green' they are, whether they happen to use recycled paper or simply offer visits to natural features. The industry, however, is beginning to realise the value of re-appraising their 'product' in the light of the more discerning tourist and consumer, who is prepared to exercise choice in how and with whom he or she decides to holiday. The President of American Express Travel Related Services, Roger Ballon, recently addressed over 1,000 delegates to the International Travel Industry Expo, held in Las Vegas, calling on them to take a lead on tourism and the environment: 'We know the 1990s is the decade of the environment. What an opportunity for the travel and tourism industry. The environment is our best product . . . If we do the right thing we can enhance our reputation by giving something back to the earth.' The creation

The Way Forward 77

of the World Travel and Trade Council in 1990 (see Appendix for details) was intended as a platform for promoting a more responsible attitude towards environmental issues. Mr Ballon has expressed optimism that these concerns can be harnessed to build a better, more healthy environment, allied to a dynamic industry. How might the tour operators translate these good intentions into actions? One way might be for them to use their economic clout to influence planners and governments to raise the standard of developments in holiday destinations, so as to ensure that the holiday and natural environment can co-exist and that policies are put in place which encourage a truly sustainable development approach.

And what of the tourist industry itself? Undoubtedly raising the standards within the industry and educating those involved in the business to appreciate the consequences of particular actions is crucial. There are encouraging signs with the proliferation of small companies offering alternative and cultural holidays and the tougher stance taken by several larger operators in the quality and type of holiday offered. The industry has shown itself reasonably willing to cooperate with environmental initiatives where the issues are straightforward and clear, such as the loggerhead turtles in Turkey. The German equivalent of ABTA issues leaflets to all its operator members which offer environmental tips to the tourist. Films have been shown on charter flights from Germany to destination countries, particularly in the Third World, which aim to educate and inform prospective tourists and travellers what to expect and how to behave as a 'guest' in the destination. These initiatives should become more widespread. (The Ark Trust are currently putting together an equivalent UK campaign.)

Within the industry the use of advertising and other media to promote tourism is full of stereotyped images of 'paradise, golden sands, friendly natives and unspoilt villages'. Probably more than any other industry the consumer (tourist) is influenced by the producers (tour operators) into believing they are making their dreams come true. Tour operators must therefore accept more responsibility and be scrupulously honest in their promotions.

Unfortunately there still appears to be no central initiative or direction within the industry itself. While ABTA now profess to an interest in green travel matters, which is good news, nothing tangible is yet on the cards. No set guidelines for their members concerning the impact of tourism on the environment yet exist. Meanwhile, the Association of Independent Tour Operators has taken a joint initiative with Green Flag International to promote sustainable tourism and is encouraging its members to consider the issues (see Chapter 12 and Appendix).

The Green Flag International initiative is potentially one of the most important developments in encouraging 'conservationists to work with the travel industry in an attempt to improve the environmental quality of holidays and to carry out specific projects at tourism destinations'. Over twenty-five tour operators are now members, including such well-known companies as Saga Holidays, and Cox and Kings, together with the national tourist offices of Canada, USSR, Malta, Mexico, St Vincent and the Grenadines, and Switzerland. A programme of projects and advice has been undertaken throughout Europe, though GFI are keen to point out that local communities are closely involved and that such projects are aimed to benefit those communities as well as the environment. Dick Sisman, GFI's energetic and enthusiastic chairman, is at pains to point out travellers' increasing interest in the environment. He feels that this interest and concern is not yet catered for by the tourist industry. A recent survey in the Netherlands highlighted this demand. The results of their survey make interesting reading:

(1) Less than 30 per cent of those responding showed no interest in environmentally conscious forms of tourism. This group were 'not interested' in any form of nature-orientated recreation or marketing.

(2) More than 60 per cent described themselves as nature lovers or open air holiday-goers. This major share of the market is interested in general nature-orientated activities such as exploring an area by foot or bike, and making long beach or nature walks. There was a limited degree of awareness about environmental issues

and a demand for tourism which satisfies a developing environmental consciousness.

(3) The third group of 10 per cent (representing proportionately the equivalent of more than two million tourists from the UK market) attached significant value to holidays which provided a high degree of nature-orientated activity – e.g. conducted forest walks, nature lectures, observing flora, fauna, and so on. This group is prepared to base its holiday choice on areas providing nature parks, historical parks and gardens, bird sanctuaries, cultural and historical monuments.

The Dutch have shown that as many as 28 per cent of all travellers form a potential customer base for green tourism. If this survey figure were directly translated into UK terms, it would represent something like six million travellers. The obvious conclusion to draw from this survey is that there is a keen demand for this type of holiday – a demand which is not currently being met. A recent survey of self-catering holidaymakers found that 70 per cent believed that environmental considerations will govern their future holiday choice.[8] Perhaps, in the end, the commercial good sense of 'going green' will change the situation within the tour-operating industry.

Despite the lack of any structured, centralised approach, individual tour operators are responding, in common with many other businesses, to environmental issues. Thomson Holidays for example, Britain's largest tour operator, recently announced a radical new departure from its traditional package holidays by opting for new destinations and presenting them in a way defined as more 'environmentally-friendly'. They display an increased awareness of local issues and are working locally to avoid potential conflicts between host nations and their own developments. They also intend to 'environmentally audit' hotels used at the various destinations. Whether this represents a fundamental change or a clever marketing ploy remains to be seen, but it sets a trend for other large tour operators to follow. There are over 600 tour operators in the UK market, many of them are small-scale outfits who have happily espoused the whole philosophy of sustainable tourism. Unfortunately, although their number is growing, they make up a minority of the number of holidays taken abroad each year. The large tour operators – such as the Thomson Group and Owners Abroad, still have enormous influence on what is available to the tourist. Because they have gone in the past for the mass market, the bulk of holidays have traditionally been low-cost high-turnover, directly contributing to the mass tourism problem. Whether other companies will follow the Thomson initiative in changing their approach to the package holiday will be one of the interesting features of the 1990s tourist industry. What Thomson's have in effect tried to do is to break up what appears to be an enormous tour operator into a number of smaller outfits each offering a range of holidays on a more human scale. They are trying to make the best of economies of scale by combining it with individual treatment.

Other encouraging initiatives include an agreement in 1991 between the International Federation of Tour Operators (IFTO) and the Balearic Islands Tourist Council (BITC) to work together on an in-depth two-year tourism project intended as a model which IFTO could use worldwide to help create and promote tourism that strikes a successful balance between social, economic and ecological demands. The study, on the island of Majorca, will aim to identify the maximum sustainable tourism flows. Majorca is an interesting place to undertake such a project, given its role in the Mediterranean tourist boom with associated costs and benefits. Martin Brackenbury, president of IFTO and chairman of the UK's influential Tour Operators Study Group, admits that the uncontrollable growth of tourism in the future will be unacceptable. 'We intend to use the study to demonstrate to other host countries means by which they can protect and enhance the future of their tourism industries,' he has said.

The very nature of the tours offered by many small-scale operators demonstrates an understanding of the impact tourism may have and a commitment to do something about it where possible. The limited numbers on each tour ensures minimal congestion, litter and disturbance of local ways of life, while the use of local guides, provisions and accommodation maximises the contribution to the local economy. Perhaps the most important measure and one which could easily be adopted by the large tour operators is the educating and informing of holidaymakers. The best of the specialist operators place high priority on the education of clients, both pre-departure and during the holiday. Extensive information is often contained in their brochures and special information packs are given, outlining local customs and culture, appropriate behaviour, environmental impacts of tourism at the destination, suggested reading and so on. Such information is regarded as an integral part of the holiday experience and is often supplemented by experienced and qualified guides and opportunities to meet local people.

Small specialist independent tour operators are leading the way. Many of the small operators working in the field of eco-tourism make significant direct donations to conservation and research projects. Eco Safaris for example are managing a National Park in Zambia with tourist money going directly into conservation projects. They have always had a company policy on conservation and green issues and these have governed both the form of their tours and the destinations they offer. Exodus travel supports the award-winning Annapurna Conservation Area Project, and Twickers World, Cygnus, Snail's Pace and many others cooperate with national and international conservation bodies such as the World Wide Fund for Nature. Such cooperation may take the form of financial contributions, on-site research, or lobbying governments – whatever is required. Traveller's Tree, for example, has lobbied both the Director of Forestry and the Prime Minister of Dominica concerning environmental issues in that country. Another good project is the ploughing of company profits into Third World charities; an excellent scheme is operated by North South Travel, a travel agent operated as a charitable trust specialising in travel to the Third World.

In addition to donations, many small operators are involved in practical conservation, often encouraging tourist participation to increase awareness. Friends of the Ionian operate a series of beach cleans throughout the season in which operators such as Greek Islands Club, holidaymakers and locals are encouraged to participate. Transglobal use a specially designed 'Nile friendly' cruiser and have instigated The Nile Initiative, an annual week-long clean-up operation. While it would take a lot longer than a week a year to clean up the Nile, these small steps help raise awareness among holidaymakers and locals.

Travel agents have a role to play in the way they promote and sell particular holidays. They have been guilty in the past of having too close an association with the tour operators and promoting only certain holidays; their role in the promotion of mass tourism should not be underestimated. In Britain we have a fairly sophisticated travel trade offering a diverse range of products. Unfortunately, however, travel agents are not always aware of the range of holidays available, nor are they knowledgeable about the destinations in the main holiday brochures and they are certainly not well-briefed on the impact of tourism on destinations worldwide. Education in this field, in order to serve the 'good tourist' is an essential step forward. Undoubtedly 'good travel agents' will help promote and influence 'good tourism'. There are encouraging signs within the industry itself that the tour operators and their associates recognise that tourism and the environment (both cultural and physical) are indivisible. By judicious management and partnership, the tourism industry can play a critical role in enhancing both aspects.

Transport

Getting from your home to your holiday destination is all part and parcel of the tourist's experience. To many, the actual travel or touring is much, or even all, of the enjoyment of the holiday, while for others it is a necessary chore in order to reach the chosen spot. You may be an avid ocean-going cruise passenger, or an enthusiastic Inter-railer; a leisurely car driver touring the Loire châteaux, or a cyclist doing the Dutch canals; a sun-worshipper wanting to reach your Mediterranean hotel and beach as quickly as possible, or a long-haul adventurer seeking to 'experience' the Orient; or simply a day-tripper to a British seaside resort. Whatever you are, the image of being a tourist or traveller inevitably conjures up some form of transport. And, of course, the more who travel, the more the requirement for the various methods

The Way Forward 83

of transport. But the very transport that brings us access to all parts of the globe is a mixed blessing in terms of the impact it may have on the environment and culture it takes us to see. The sheer scale of numbers involved in being moved round the earth requires a huge amount of resources. Roads and airports have to be built and that in turn leads to vehicle emissions and a detrimental effect on air quality.

Air travel, the single biggest means of travel for international travellers leaving the UK, is set to double in terms of passenger numbers by the early years of the next century, to around two billion passengers worldwide each year. Despite recent blips as a result of recession and war, passenger traffic growth curves soon recover once crises are over, as happened in the 1970s and 1980s. How are major airports worldwide, currently at saturation point, going to cope? How long will the air traffic control delays be in the next century? One answer might be to build more airports and runways, but the resistance among evironmentalists and those living near proposed airports has grown enormously of late. A more realistic answer is to search for more efficient transport, either in the form of bigger aircraft, or an altogether alternative means of transport. The development of a reliable high-speed rail network in Europe, and the Channel Tunnel linking Britain to that network, are signs of a realistic alternative within western Europe, which can readily compete with the airlines for journeys under 450 miles, in terms of time and efficiency, from both an economic and environmental point of view.

Pollution by aircraft is increasingly under scrutiny. 'Hostility towards air travel could suddenly reach levels currently directed against sections of the nuclear and chemical industry', was the warning given by Dr Michael Grubb, a climate adviser to the UN, when speaking at an international aviation conference in Geneva in 1990. The problems are ones of noise, fuel demands and noxious emissions. The airlines themselves are aware of the criticisms.

In response to a 1990 survey, Air Canada, Air France, British Airways, Japan Airlines, Lufthansa, Quantas and Singapore Airlines all stated that they have formal policies on environmental protection, including pollution and waste disposal controls, noise abatement and fuel consumption efficiency measures. All the airlines also documented their commitment to recycling. Three of the airlines, BA, Lufthansa and Air France, have recently appointed environmental executives to develop and implement such policies. They deny that the appointments are a reaction to outside pressures or consumer 'greening' but it is interesting to note that BA's directive for its environmental executive, for example, includes 'ensuring the airline is not vulnerable to external criticism' and 'reassuring the public'. One important but little-known initiative on the part of British Airways was started up by one of their engineers in 1983. Spotting that BA, with its wide network of routes throughout the world, was in a unique position to help conservation organisations, Rod Hall persuaded the airline to agree to an 'Assisting Nature Conservation Programme', whose Principal Coordinator he now is and which is now flourishing more than ever. This involves the transportation of rare species, in cooperation with four internationally respected conservation organisations, from one part of the globe to another free of charge. Re-introduction schemes for endangered species are greatly helped in this way, and illegally imported animals are restored by BA to their country of origin. Equipment and, on occasion, personnel are also offered free passage, so assisting habitat study and conservation education. Because the programme works mainly on a system of space availability, it does not compete with income-producing passengers and is therefore an idea which other airlines could (and should) easily adopt.

The British Airports Authority has recently introduced a corporate advertising campaign which features the internationally respected naturalist Gerald Durrell reporting that 'jumbos can co-exist with a herd of Roe deer'. The use of environmental issues in the advertising and marketing of airlines is certainly a recent development and must be linked to consumer interest. Many airlines maintain they have been 'green' for many years although Air Canada, the world's only non-smoking airline, is ready to admit that it is predominantly recent public awareness of ecological issues that has affected the airline's corporate policy.

Many airline environmental policies exist to conform to international and national legislation rather than because of any deep-rooted commitment to protecting the environment. For example both IATA (International Air Transport Association) and ICAO (International Civil Aviation Organisation) provide guidelines and restrictions upon noise and pollution levels and many airports restrict incoming flights accordingly. PanAm recently stated that, '. . . with the possible exception of legislation in Germany, we remain . . . certain that environmental issues will not have a significant impact on airline operations in the near future'.[9] However IATA has now set up an environmental task-force to co-ordinate the airline industry's response to the issues. When assessing an airline's commitment to the environment it is useful to consider the question of fuel efficiency because there is a strong economic motive behind this. Most airlines though *are* working, for whatever reason, to increase their airline's efficiency.

Aeroplanes are certainly consumers of large quantities of fuel, but they also represent better fuel efficiency in terms of numbers of passenger miles per litre than cars. For example, a charter plane full of tourists from Britain to the Mediterranean is likely to be more fuel-efficient than if those same tourists drove their cars there. As major players in the tourism industry, the international airlines have a significant role to play in encouraging the 'good tourist' directly, through their own operational practice, and indirectly, through the provision of information to tourists who travel with them, particularly to Third World countries. The tourist in turn can exert influence by choosing those airlines which reflect this approach.

For many people the car is essential, and they are loath to part with it, even (or especially!) when on holiday. The problems of traffic congestion, familiar to most of us in large cities, have been transferred to many holiday resorts and destinations. The convenience of the motor car is undoubted, but when on holiday, who wants to sit in a traffic jam or constantly worry about a parking space? The answer lies in a variety of directions. Obviously the need to provide realistic public transport is critical. Many countries outside Britain view their public transport systems as a service and subsidise them accordingly, rather than seeking to make them self-financing. You will usually find a greater variety of public transport options than at home. Why not try them more often? You'll find it much cheaper than taxis or hire cars, and it gives you an opportunity to meet the locals. Alternatively, walking or cycling can be a real eye-opener, as it tends to make you more accessible to your surroundings and to the people who live there.

Of course there is much that the industry and governments can do in terms of better management of transport. The 'Tourism and the Environment' Task Force report, produced for the English Tourist Board, suggested that good management can improve the transport situation in many problem areas by strategies such as:

- directing visitor traffic away from sensitive areas
- suitably located car parks associated with public transport, i.e. 'Park and Ride'
- integrating public transport and developing facilities such as cycle hire
- closing roads to all but local traffic, walkers or cyclists (such tactics are used in the Black Forest in Germany).

In Windsor, for example, one fifth of all visitors to the Royal Borough arrive by coach. A recently constructed coach park and visitor reception area, including restaurant, tourist information centre and associated facilities, has been designed to take coaches away from the town centre whilst at the same time a special route has been developed so that visitors can walk from the coach park to the town centre.

Such initiatives represent a realistic way of reducing the impact of vehicles on environments not built to cope with the increasing levels of traffic demand.

As already mentioned, the prospects of travel to Europe by train are improving. The Inter-Rail pass, giving unlimited travel to virtually every country in Europe, has recently been extended to allow those over twenty-six to take advantage of it. It also offers a superb opportunity to explore the cities, cultural high spots and rural backwaters of Europe using the most environmentally efficient and cost-effective method of transport (see the latest edition of *Europe by Train* by Katie Wood, published in paperback by Fontana).

The Environmental Transport Association is a non-profit-making organisation. It was launched in 1990 to promote environmentally friendly forms of transport and to protect our environment and heritage from the damaging effects of transport (see Appendix for further details). ETA has produced information to encourage responsible travel by tourists.

In summary, transport has a major impact in creating a demand for more resources to build roads and airports. Many initiatives are being taken to improve fuel efficiency, but this is as much to do with controlling operating costs as anything else. Tourism cannot be divorced from transport issues. Accordingly only when transport issues in general move on to a green level – improved public transport, more efficient fuel use – will there be any impact on tourist travel.

Accommodation

The hotel industry often appears to embody many of the worst aspects of mass tourism and is one of the most criticised features of the industry. To many people, hotels represent the worst of the horrors of inappropriate development, with generally remote ownership and low quality, usually obtrusive buildings, which contribute little to the welfare or environmental well-being of the locality and generally detract from the overall character of a resort or destination. The worst examples are found in Third World countries where the demands of big hotels may even deprive the local community of such essentials to life as water and power, as has happened in Goa and Tunisia. In such situations the hotel industry, having invested heavily and with the support of government, will seek an economic return on its investment, perhaps without having fully considered the impact of its operations.

The hotel industry would argue that, without large hotels offering reasonably inexpensive accommodation, there would be no mass tourism. They might also argue that they do provide considerable employment in holiday areas and generate significant amounts of tax revenue for both local and national government. There are also many examples of the hotel and accommodation industry renovating and restoring old buildings to serve a very suitable new use as living quarters for tourists. The ebullient Dr Henry Fraser, President of the Barbados National Trust, is dedicated to the conservation of the many fine old buildings in Barbados and believes that sensitive restoration for tourist accommodation is one of the ways to save otherwise threatened plantation houses and mansions.

Of course bed and breakfast, as offered throughout the UK and increasingly in other countries, represents an ideal way of accommodating visitors in a suitable way, building bridges between guest and host. In particular, it gives local people a stake in tourism, both personally and financially, by ensuring that the revenues go straight into the local pockets, which can in turn help to support local services. Bed and breakfast not only provides an important source of revenue, often critical in rural areas, but also now serves as one of the outstanding features of the tourist experience in Britain. Good practice in the provision of accommodation for tourists is essential to ensuring that tourism is a positive experience for any community. The English Tourist Board have recently produced such a good practice guide called 'Developing Rural Accommodation', together with a 'green tourism manual' for both tourism operators and developers. Such an approach is a positive way forward.

And what of the large multinational hotel groups? How are they responding to the demand for a more human face to the tourism industry? The Boston-based ITT Sheraton Corporation is a global network of nearly 900 hotels, inns and resorts, operating in some sixty-five countries – the leading hotel company in terms of international coverage. Sheraton has recently introduced a 'Going Green' scheme for its guests, employees and the local community. As one of the main programme elements in the 'Going Green' campaign, it introduced an 'Optional Dollar' initiative which invited all guests at Sheraton hotels in Africa and the Indian Ocean to add one dollar onto their final bill when checking out. Sheraton then matches this amount in local currency. In the last quarter of 1990, some US $158,000 was raised for local community and wildlife projects identified by the World Society for the Protection of Animals. One of these projects included the Yankari National Park in Nigeria, where Sheraton donated two land rovers and anti-poaching equipment. In 1991 Sheraton undertook an environmental review of the fundamental infrastructure of each hotel. The specific conservation guidelines include:

- the reduction in external pollution
- the reduction of dependence on non-renewable resources together with improved energy conservation
- improvement of the environment
- minimising the creation of waste products and encouraging a recycling programme.

In Zaire, the Sheraton Karavia Hotel is working with World Vision, local specialists in rural projects, to provide drinking water for Kwana, a village in Shaba province. Such initiatives recognise a new approach by a large company such as Sheraton. In many ways they also illustrate the influence of the tourist and traveller as a consumer to persuade multinational organisations to change their policies. Mike Prager, Sheraton's Vice President and Director of Marketing in Africa and the Indian Ocean has commented on the 'growing awareness of the need for responsible tourism which we hope to increasingly cultivate among travellers. Our long-term aim is to strengthen our resources and expertise in this area within our industry.' Clearly Sheraton has seen the need (and the advantages in terms of marketing) for reviewing its practices.

Other hotel groups, such as Hilton International and Marriot, have made noises about environmental policies, whilst the Inter-Continental Hotels Group recently announced its own programme of environmental initiatives, mainly concerned with recycling waste paper and using environmentally friendly products in housekeeping duties.

Overall there is no doubt that the hotel industry is becoming aware of the importance of its image in relation to the environment and to the views of its customers. As yet the translation of this recognition into better practice is limited, with the notable exception of Sheraton. However the stirrings are there and, with the right incentives allied to corporate planning, control and monitoring, the accommodation industry can play a positive part in promoting 'good tourism'.

The Good News – so far

'Tourism must underpin conservation. It has got to.'

David Bellamy.

In earlier chapters we have discussed the very real problems of tourism, together with potential solutions to those problems and ways of increasing the real benefits. But what of the future? What hopeful signs are there for us to grasp and use as examples to others? Are the 1990s going to be, as widely predicted, the decade of the environment? How are those who have to make a living out of the tourist industry going to reconcile the apparently contradictory demands?

There does appear to have been a fundamental recognition amongst governments, tour operators, travel agents and tourists themselves that the tourist industry cannot exist in isolation from major environmental issues, nor can it blithely ignore the dilemmas caused by reaching the limits of destinations in terms of visitor-carrying capacity. Conferences, policy papers, codes of practice and good intentions abound, but how can they be translated into practice? What examples are there to show that the issues can be tackled and solved?

The good news is that there are plenty of examples of good practice. Starting at home first, the UK government's 'Tourism and the Environment Task Force' launched its report in 1991 following extensive consultation within the body of organisations and interests dealing with tourism to and within the UK (though not UK tourists going abroad). The Task Force Report set out principles for the balanced development of sustainable tourism. These are:

– The environment has an intrinsic value which outweighs its value as a tourism asset. Its enjoyment by future generations and its long term survival must not be prejudiced by short term considerations.
– Tourism should be recognised as a positive activity with the potential to benefit the community and the place as

well as the visitor.
- The relationship between tourism and the environment must be managed so that the environment is sustainable in the long term. Tourism must not be allowed to damage the resource, prejudice its future enjoyment or bring unacceptable impacts.
- Tourism activities and developments should respect the scale, nature and character of the place in which they are sited.
- In any location, harmony must be sought between the needs of the visitor, the place and the host community.
- In a dynamic world some change is inevitable and change can often be beneficial. Adaptation to change, however, should not be at the expense of any of these principles.
- The tourism industry, local authorities and environmental agencies all have a duty to respect the above principles and to work together to achieve their practical realisation.[10]

The next step will be for the English Tourist Board to implement the proposals, which it plans to do by encouraging partnership between public and private bodies and by using local area initiatives to plan and manage tourism in specific areas. Further initiatives include an enhanced Green Tourism Award for England for Excellence, which will publicise the most successful demonstrations of the balance between commercial success and environmental awareness.

The government is also belatedly acting to clean up its beaches and improve the quality of its bathing waters. It has announced a £1.4 million investment programme designed to bring all UK bathing waters up to EEC standards by the end of the century. A further £1.5 billion is being spent on improving sewage treatment near coastal and estuarial discharge outlets. Bathing water standards are monitored under the Blue Flag Scheme, sponsored by the EC and supported in Britain by the Tidy Britain Group and the English Tourist Board. The government is also taking action on the litter problem, by giving £3 million in grant aid to the Tidy Britain Group to fund twenty-seven clean-up projects.

Elsewhere, countries as diverse as New Zealand, Singapore and the Seychelles are thinking – and acting – seriously about how to ensure that they maintain the balance. In New Zealand, the major attractions for tourists are those natural scenic wonders and unique features such as the Rotorua geysers, the Milford Sound and the glow-worm caves of Waitomo. In the 1970s and 1980s, visitor numbers to these attractions and their associated activities grew enormously, presenting challenges to those charged with protecting these sites. By recognising the precedence of sustainable use over exploitation, and by implementing appropriate visitor-management techniques, the attractions have been protected for the benefit of tourists, operators and the local residents.

In Singapore the pace of change in the last twenty years has been enormous. New buildings and developments have sprung up in a very short space of time, replacing the ethnic Chinese, Malayan and colonial buildings that formerly gave the city such a unique style. The city was dedicated to material advancement through tourism (particularly attracting stop-over visitors on long-haul flights), finance and industry. However there was a feeling that the heart had been ripped out of the island; a soulless, cosmopolitan city had taken over, which could be anywhere in the world. Happily the Singapore government has recognised how important its rich cultural mix and history was in attracting tourists. Now the push is on to restore and conserve areas such as Chinatown. The change quite clearly has been brought about by pressure from the tourism department who acknowledge the interests of tourists in this aspect of Singapore.[11]

In the Seychelles the government has recognised the reality of modern economic life. The pressures to earn foreign revenue mean that the islands need visitors, as tourism is the biggest contributor to the economy. Nonetheless there is a clearly stated policy recognising the capacity of islands to cope with visitors and the appropriate type and location of development. A tourist 'ceiling' has been set for the mid 1990s, based on estimates of what the islands can sustain without social or environmental damage.

On some islands tourists are allowed to wander, whilst on others they are strictly limited in terms of when and where they can go, or indeed discouraged altogether.

All this demonstrates that governments are now at least thinking about the impact of tourism, and the proliferation of 'draft policy papers' and 'outline proposals' to improve matters regarding tourism in the last year has been quite remarkable. Of course there is a considerable amount of jumping on the bandwagon and many of the current statements have yet to be translated into action, but at least the issue is on the agenda and being talked about.

As might be expected, the northern European and Scandinavian countries have had tourist policies which espouse environmental values for some time. Finland, Austria, Switzerland, Norway and Germany all have national tourism policies which reflect their sustainable development approach, and tourism developments are assessed for their environmental friendliness before being passed. For example, the Finnish Tourist Board favours purpose-built tourist complexes which do not 'pollute architecturally', and ensure adequate sanitary facilities, fuel supplies and so on. These countries also provide extensive information in their brochures and promotional material, encouraging their visitors to be well informed, arriving in their country with a better understanding of it and its people and culture. It may be argued however that these countries do not receive the high volume of visitors that creates many of the problems green tourism sets to solve.

Spain, the destination where most UK residents travelling overseas go to on holiday, has suffered all the environmental difficulties associated with mass tourism. The Spanish government is beginning to recognise the importance of sustainable tourism development, largely because inflation and changing holiday trends are leaving the Spanish resorts facing a long-term slump. Its tourist industry is now concentrating on the smaller-scale richer end of the market, and the face-lift that this requires can only be good for the environment. Recent initiatives include the creation of a Council for the Environmental Protection of the Mallorcan coastline and the appointment of an Ecology Counsellor to spearhead Benidorm's Green Campaign. Both central and regional governments are highlighting improved standards in environmental issues to encourage more public awareness and participation. Quality rather than quantity is now the philosophy and only controlled building of a limited number of high quality properties will be permitted in tourist areas such as the major cities, the southern Costa de la Luz, and the relatively undeveloped northern coastline. The 1988 Shores Act banned any construction within a strip of between 100 and 700 metres of the waterfront and gave local authorities powers to order the demolition of buildings found to lack planning permission. Owners who possess some but not all of the necessary building documentation, of whom there are many in Spain, will be required by the local authority to wait for thirty years before any such proposals will be considered for approval, during which time they will in effect be unable to sell their properties.

In Catalonia, home of the Costa Brava and the location of many of the early package deals for British tourists, the emphasis has moved onto attracting tourists into the countryside away from the coast. Walking and cycling tours, farmhouse accommodation and highlighting the distinctive Catalan culture are all part of this change. Even in the Balearic Islands the aim is to attract more people away from the coast to sample the countryside and the farmhouse accommodation now being set up.

It is not only at government level that such initiatives are seen. The industry itself is discovering new 'greener' ways of operating that are proving successful. For example, in the Australian state of Tasmania the move towards a more cultural approach to tourism is evident in such projects as the Colonial Accommodation scheme which arranges 'colonial-stay' holidays, providing accommodation in private homes instead of the usual motels. The scheme grows in popularity every year and tourists gain a deeper insight into Tasmanian life through this genuine contact with the host population. Such schemes are also springing up in Britain. Here the Rural Tourism Development Project works with local villagers to develop small-scale rural tourism schemes, where accommodation is provided by the local villagers and the emphasis is upon a host/guest relationship, rather than one of producer/consumer. Tourism has been developed in this way in the Brit Valley in Dorset, Blakeney in Gloucestershire and Dulverton on Exmoor.

In Florida there has been much debate about the massive impact of development, much of it tourist-related. Here responsible commercial interests are showing that they can play a major part in maintaining the natural stability of the environment in which they operate. The Disney Corporation has always had one eye on the way its activities fitted into the natural world, from which Walt Disney drew much inspiration. The original Disney World was developed using plans drawn up by some of the USA's premier conservationists, leaving large areas undeveloped within the theme park. All matters of infrastructure such as water, sewage, energy and building specifications are all undertaken in a state-of-the-art manner aimed at eliminating detrimental impact on the local natural environment. The key, according to the head of environmental affairs, is to use Disney's massive purchasing power to ensure that suppliers can use or provide recycled water, paper and materials.[12]

The Bellerive Foundation's Alp Action project is another excellent example of an organisation whose concern over the negative impact of tourism on a fragile environment has prompted direct and imaginative action. Three schemes for which they are currently requesting international sponsorship are: the setting up of a prototype tourist reserve at Lauzière in Switzerland which will aim to integrate tourism and conservation in the development of a ski resort; the distribution of a code of ethics credit card advocating good behaviour in the mountains to all tourists visiting the Alps and which could entitle visitors to various discounts on such things as public transport and sports and leisure facilities; the publication of a handbook for mountain-users, called the *Roof of Europe*, which outlines conservation problems, details protected areas and endangered species and presents tips for travellers and tourists. The idea behind the Alp Action project and other similar programmes is to encourage a partnership between business and tourism in order to foster a caring, responsible attitude to the environment. The signs so far are encouraging but there have been no major, large-scale changes as yet and such decisive action is urgently required.

There have been several initiatives pioneered in North America to rehabilitate degraded environments such as old docklands and wharves together with associated heavy industrial sites for tourism and leisure use. Particularly good examples can be found in Montreal and San Francisco and have been mirrored in the UK by 'smokestack' tourism in places like Bradford, Manchester, Glasgow and Liverpool which have repackaged their historic industrial heritage to attract visitors.

The Annapurna Conservation Area Project was founded in Nepal in 1986 in response to problems caused by tourists trekking in the Himalayas. Tourism is Nepal's major foreign earner and the main form of tourism is trekking. The project was set up, funded by the World Wide Fund for Nature and the German Alpine Club, to encourage local people and tourists to cooperate in developing a more sustainable form of trekking tourism and alerting tourists to the problems of deforestation, erosion, and pollution of water courses and sources of drinking water.

The high public profile of such schemes as the Blue Flag scheme for beaches has helped capture the public's imagination and support. The recently formed Green Flag International aims to do something similar by offering specialist advice to tour operators about the environmental quality of holidays and ultimately to help the industry have a higher regard for the conservation and enhancement of the coast, countryside and communities at tourist destinations.

The rapid growth of influence of organisations such as Tourism Concern is another indication of the public awareness of the issues and willingness to get involved in doing something about it.

For some countries, of course, like Bhutan, the limitation of tourism to very low numbers, may be appropriate. Throughout the current debate on the impact of tourism there are many who feel that the problems are insurmountable and that the projected vast increase in tourist numbers will simply swamp attempts to reconcile those problems. That is not our view. There are enough examples, we feel, to show that there are potential solutions which can be implemented, provided the will is there. Solutions will involve a variety of different approaches, yet there must be a closer relationship between the tourist industry, the host community, the tourists themselves and local and national governments. And these relationships must be forged across international boundaries because the scale of the industry is such as to require multinational solutions.

The role of good practice and a high standard of tourist management is also critical. Of course this will involve training and resources. But this must be considered as a necessary investment. Tourism has to be viewed long term to encourage sustainable development policies, policies which will encourage viewing tourists not as a 'goldrush' to be mined as efficiently as possible until the seam disappears, but rather as a crop which can be harvested year-in, year-out in perpetuity, provided it is nurtured, cultivated and properly managed, finding a balanced niche in the economic, social and environmental order of things.

Chapter 6

The Good Tourist

The Influence of the Tourist

Having considered the tourist industry and the influence of governments and other organisations on tourism, what about the tourists themselves? Does the individual have the influence necessary to effect real change, given the multi-billion dollar nature of the industry?

The aim of this book is to attempt to guide the individual and offer some advice and tips on being a 'good tourist'. Much of this is, of course, presumptuous in assuming that one type of tourism or one type of behaviour or one course of action is better than another. In the end, however, we all have to make choices and the more informed we can be, the better. By exerting their consumer power, tourists and holidaymakers *can* have enormous influence on the market, as has been evident recently in traditional Mediterranean resorts. It's not as if leisure time is decreasing; quite the reverse. And while the current blip in the economy has caused a hiccup, it has not stopped the overall holiday market continuing to grow. Although bookings for package holidays decreased in 1991, the overall number of UK residents taking holidays increased compared to the previous year. People have again voted with their wallets and are beginning to want something different.

There's no shortage of advice on how to behave as a tourist. There have been a number of guidelines, credos, tips, checklists and so on produced recently by a variety of organisations offering tourists advice on how to be 'better', more thoughtful visitors to foreign lands. Some are short and punchy one-liners, primarily highlighting the environmental problems of tourism. Among these are the Countryside Commission's 'Guide for the Green Tourist', the English Tourist Board's '20 tips for tourists', 'A Traveller's Guide to Green Tourism' produced by Green Flag International,

Ark's guide, and 'Guidelines for the Mediterranean Tourist', produced by the Earthwatch organisation in conjunction with Greenpeace and Friends of the Earth. On a deeper level are the guidelines produced by organisations concerned with the effects of tourism on Third World destinations. These include the 'Guide Notes for responsible Travel – a credo for the caring traveller' produced by the Centre for Advancement of Responsible Travel, together with the guidelines issued by the Tourism with Insight International Working Group, aimed not only at the tourist but also the travel industry. These guidelines consider not only environmental matters but also social, economic and cultural aspects of tourism, particularly in under-developed countries. (For addresses and further details, see Appendix).

Such guidelines raise many of the issues of tourism, and while as tourists we don't constantly want to be bombarded by 'do's and don'ts', it *is* something the industry itself might consider developing and promoting. In much the same way as most of us are familiar with the main message of the 'Country Code', so might a 'Tourist Code' serve a similar purpose.

Do you *really* want to be a tourist?

The first thing to ask yourself is: do I really want to go on holiday? As discussed in Chapter 2 people go on holiday for all sorts of different reasons: often for reasons that they won't admit even to themselves. There are many people who go on holiday not really because they want to, but because it's the thing 'everyone else is doing at this time of year'. But there's no great competition to go travelling to foreign countries! Certainly many do it, and there's a part in all of us that harbours romantic notions of the 'great journeys'. But what's the point of contributing to the tourist load if your holiday is only going to leave you feeling worse than before because a careless choice landed you with a trip that only aggravated the frayed nerves you were trying to soothe. And how many foreign holidays, even packages are from start to end genuinely relaxing? If the motivating factor behind the decision to holiday is to relax and unwind after the stress and strain of a year at work, a hectic week of sightseeing in high temperatures, a forty-eight-hour delay at the airport, or hours spent waiting in holiday traffic jams with squabbling children in the back seat, is hardly likely to achieve that objective. Faced with these kinds of holiday hassles, it's little wonder that we've no time or energy to immerse ourselves in new surroundings and a different way of life, or that we are short of patience when it comes to contact with local people!

If you feel like that, don't travel. Consider having a holiday at home, exploring the immediate vicinity of your neighbouring towns, city or countryside, as if you were a tourist. Many of us rarely look at our own areas and appreciate their true value. Alternatively, you might simply relax! If the intention of a holiday is to have a change from the normal routine and to pursue different things from everyday life, is it entirely necessary to leave home to do it? If, however, you do decide to go away on holiday, try to think about *exactly* why you want to go; where you want to go; who you want to go away with and who you want to meet on holiday. Once you have decided these things, then decide on the most appropriate type of holiday for you, together with how to organise the whole thing. There are many hundreds of different types of holidays available, making it extremely difficult to choose an appropriate one that suits your requirements. In general, foreign holidays are either package, all-inclusive deals, or independently organised by the travellers themselves, though there are variations and hybrids of the two. Holidays can be either static – in one or two places – or touring round regions or countries. They can range from resort-based holidays through caravan and hiking tours, to activity holidays – walking, skiing, golfing, fishing etc; from the hard-sell of commercial resorts to the alternative approach of 'soft' tourism, with a gentler approach to the destinations and its people and resources; from ecotourism, visiting national parks, mountains and wilderness areas, to the cultural highlights of Europe or India; from educational study groups and voluntary work to the sex shops of the Far East. Tourism reflects life in the late-twentieth century and 'all life is there'. No wonder the tourist gets confused and often opts for what he and his travel agent knows and recognises rather than what he really wants.

Different types of holidays have a different impact on the environment or country in which they operate. The 'stay at home' option is therefore very attractive in terms of having a minimal negative impact as a tourist – you can't get more minimal than staying at home and not being a tourist – though this is a bit extreme! Alternative and responsible tourism is trying to reconcile the desire to be a tourist with the need to avoid having a negative impact on the destination – or at least balancing it with a greater positive contribution. Similarly, educational holidays tend to be low on impact and aim to give the tourist a new perspective on a subject. Being a volunteer on holiday is becoming more popular. The cost is generally a lot less than conventional holidays and they make a positive contribution to the environment, either man-made or natural, in which they take place through such things as archaeological digs, helping disadvantaged members of society or practical conservation – (tree-planting; repairing eroded footpaths, wildlife census and analysis).

These types of holidays are the 'goodies' if you like of the market as far as the 'good tourist' is concerned. However not all of us have either the time or the inclination to follow such an approach in our precious leisure time, yet we may still be genuinely concerned about the type of holiday we choose. Certain outdoor holidays such as hill-walking or trekking, wildlife observation safaris and adventure trips to wilderness areas, or cultural trips to other societies, attract many people who have an empathy with nature or the societies they are visiting. But great care is required here to ensure that the activities you might want to undertake are done in a way that is compatible with the aims of 'the Good Tourist'. It is anomalous to go and view a great cultural or natural resource if you are contributing to its disappearance or destruction. Footpath erosion in the Lake District and Yellowstone National Park in the US; deforestation in Nepal, and damage to the Parthenon in Athens and the Sphinx in Cairo are all examples of this type of problem.

Most British tourists, however, go on conventional holidays package or independent. The 'green' view might be that this is the most destructive type of holiday and as it is the majority of people who are doing this there is a problem and a need to change attitudes and trends. This type of holiday caters for the movement of large numbers of people and caters for their high consumption of natural resources, all of which contribute further to overall global pollution. Many people believe they have little choice in the type of holiday available to them given the time and money available, however, as the strong green consumer movement has shown in the last couple of years, the individual *can* influence many aspects of industry and demand can influence supply. This is what's needed in the tourist industry.

Realistically not everyone is prepared to take all of the steps necessary to espouse low impact or alternative tourism, but if people who are going on more conventional holidays start taking small steps towards promoting responsible, sustainable tourism, then the tour operators *will* undoubtedly respond. There are already signs of this happening, as in Thomson Holidays' greener image and 'environment audits'. While the sincerity of this approach is yet to be fully tested, it is indicative of a trend that many other operators are likely to follow.

Environmentally Audit Yourself

There's no magic formula to becoming a 'Good Tourist'. As we've tried to convey in this book, every type of holiday or journey can have a varied impact depending on the situation or the time. Thus a package holiday to Benidorm may, in the long run, have much less of a damaging effect than a safari to a national park in East Africa, or vice-versa. The important thing is for you as a tourist or traveller to have sat down and thought about *why* you are going away and analysed your motives within the context of the experiences you are looking for and the impact on the people and places you will be travelling to. One way to do this is to 'audit' yourself and your holiday. Go through the following checklist and associated suggestions before, during and after any trip and ask yourself – are you a Good Tourist?

BEFORE YOU GO

1. *Why go on holiday?* As discussed above, the first thing to do is to question your motives. If you don't really want to go away,

don't! Relax at home and take the chance to enjoy the pleasures of your own home and region. Don't follow the crowd.

2. *Identify exactly the reasons why you're going on holiday.*
   – do you want adventure and excitement?
   – do you want to meet new people?
   – do you want to rest and relax?
   – do you want to travel to new countries?
   – do you want to experience different cultures?
   – do you want to have a good time and concentrate on a suntan and nightlife?
   – do you want to travel alone or in a group?
   – do you want to travel independently or on an all-inclusive package?
   – do you want to visit a particular region or country? If so, why?
   – do you have a special interest or reason for going on this trip? Examples might be visiting the cathedrals of Italy, birdwatching in Turkey, visiting friends or relatives, or even doing business in a particular country.
   – do you want to stay in one place or move around?

Of course, you might answer 'yes' to several of these; there are doubtless, somewhere, sun-worshipping golfers who want to tour the Andalucian monasteries, preferably on a package tour, while at the same time having a few days to pop over and see the relations on Gibraltar. On the other hand many people just want a holiday as a change from the normal routine and as long as it is a reasonably pleasant change, they'll be happy.

3. *Choose the right holiday.* Having decided exactly why you want to go on holiday, next choose the right type of holiday to suit your purposes. There's no point going off to the other side of the world for a packed fourteen-day tour of every Inca monument if what you really want is a break from your normal routine and the stressful, frenetic lifestyle most of us lead. All you'll do is replace one kind of stress with another. Many people return from such trips exhausted, disillusioned, restless and discontented with

their lifestyles, and more stressed than before they left. They are delighted to get back home to their own beds, so all's well for the first few days back, but they still need a rest and there's no more holiday till next year. So make a lifestyle travel plan. Be realistic. OK, you might want to see the Inca cities, but given your present career, young family, income etc., is this really the optimum time to go or has the fact that it's in all the Sunday supplements got a lot to do with it? Why not jot down all the places you really want to go to, along with the genuine reasons why, and roughly allocate times to the trips? Earlier retirement and longer, healthier lives mean trips to Peru, Australia, the Caribbean and the like can all take place when you're older, when the children are grown up and away, when you're retired. In the interim years read about these places; learn about what it is there you really want to go and see. When you do get there, your visit will mean so much more to you. Don't rush. The world will still be there, hopefully, in a few years' time. Don't contribute to the 'fortnight to see Europe syndrome'; take your time and plan it to suit you.

4. *Travelling out of season or to less well-known regions.* Think about travelling off-season or at either end of the high season to avoid the busiest and hence most stressful periods. Consider travelling and visiting places not as well-known as the normal. Doing it this way is often less expensive than going to the busiest places in high season. You're also not then adding to the already serious tourist congestion problem worldwide.

5. *Choose the right travel method and tour operator.* Be selective in how you travel and the travel company you use. For those of you choosing a holiday with a tour operator this might mean being more demanding than normal and asking questions that the average travel agent won't be able to answer, but write to the tour operator and the questions will be answered. Ask about tour operators' policies and practice in relation to environmental issues, such as do they themselves subscribe to waste recycling, cleaning up pollution around holiday resorts and taking part in local initiatives to maintain the balance between tourism and a

healthy environment and happy local community? Do they know how many tourists are in that resort at the height of the season? What are the local ecological and environmental issues there, and is there any conflict between tourism and these issues? If so, what are *they* doing to help and what are their views on the subject?

Tourists travelling to Third World countries might also ask what the economic benefits of tourism are to the local economy, or whether the hotels and facilities used are locally or internationally owned. Staying in a direct-benefit B&B or locally owned hotel is far better for that country (as is flying the national airline). If the travel agent or tour operator doesn't know the answer to who owns the hotel/airline, the chances are it's because he's never even considered the question before. By choosing a holiday with such an operator you are unwittingly perpetuating inappropriate tourist development. There are many small tour operators springing up who have considered such questions and who are willing to provide the answers. They may, because of their size, be more expensive than larger operators, but that may simply be a matter of scale – if consumers, and that's what tourists are, exercise their powers, such important issues will be reflected in the future price as more people choose to raise these points. There are signs that this is happening already but you should be on guard for those merely paying lip service to the values of responsible tourism.

The travel industry is so developed and sophisticated that there is very little that is not available these days and money is no longer the stumbling block it once was for travellers. Don't give up if your travel agent tells you what you want is impossible. It *will* be possible with some research and determination. Don't settle for compromises. Not now you really know what you want! There is no good reason why both the trips and the type of trips you want can't be a reality.

A good tour operator will attempt to work with the local community to ensure their operations are as environmentally and socially acceptable as possible. To summarise then, ask:

- Do they contribute to local initiatives to keep the resort in good condition?
- Do they create local employment?
- Do they use accommodation which is at least locally owned and built in the vernacular style?
- Will they try to integrate holidaymakers with locals?
- Do they inform you properly about the destination prior to travel?

6. *Advertising*. Be wary of holiday advertising that is clichéd and generalised, using phrases such as 'tropical paradise, with unspoilt beaches and friendly locals'. It may have been that way at one time, but it certainly won't be after all the bookings have gone into the computers! Try to get behind and see through the adverts. If you're looking for a quiet spot, go through all the brochures and then pick a location not featured in any of them!.

7. *Education*. Get impartial sources and find out more about your holiday destination and the people of the country before you go. (Hopefully, you'll find the individual country guides within the 'Good Tourist' series of use here.) Even if you are going for a two-week package to rest and relax, this information will give you a greater insight into the country, the people, the region or resort you are visiting and the customs and way of life there. There are numerous excellent guidebooks now available to virtually everywhere in the world. Your local library will also have a stack of information, free of charge. Contact the London-based Tourist Board of the country. It is particularly useful to do pre-departure research when visiting Third World countries with different cultures and norms. Bear in mind that the Tourist Board literature will gloss over the real problems, but, balanced with other information, it will help you.

Not all of us will be able to satisfy all the stringent conditions that apply to the green or good tourist, but every step, no matter how small, adds to the sum of the overall responsible tourism effort. Learn the language – it needn't be 'total immersion' but we all know how much we appreciate visitors' attempts to communicate and how frustrating it is all round if the visitor doesn't make the effort to learn the simplest phrases.

8. *Packing*. Take nothing in your suitcase that isn't absolutely necessary. Remove items from their packaging; in many countries waste disposal as we know it just doesn't exist and the opulent tourist can produce more waste in a fortnight than a local does in a year. So remove your toiletries, photographic goods and clothing from their polythene and cardboard, and dispose of that at home before you go. Also choose your sunscreen and shampoos carefully. By taking ones made from natural substances (e.g. Body Shop ones, and the like), you will help preserve the marine life and water purity of the area.

THE HOLIDAY ITSELF

9a. *Transport*. It goes without saying that in order to be a traveller you need transport. Most of us travel by planes, boats, trains or cars in order to get to our destinations quickly, given that holidays are often precious windows in a busy life. Consider your form of transport when travelling both to your destination and touring around. Do you really need a car? Why not try either walking, cycling or public transport? These three options are generally less stressful than the car and help you absorb the atmosphere and spirit of the destination you are holidaying in instead of polluting it! They also let you meet the locals more easily.

If you must take the car, try to use Motorail where possible, and use unleaded petrol. Check the technical state of the vehicle before setting off to ensure maximum fuel efficiency and minimum noise and carbon dioxide pollution, and drive at 65 mph, not above. Try to avoid busy motorways, mountain passes and border crossings at peak times.

If you've flown to your destination, try to use public transport links to the city centre instead of a taxi.

9b. *Accommodation*. Go for traditional, established, locally owned and run inns, hotels or B&Bs. Choose smaller establishments which aren't blots on the landscape and offer the typical food and hospitality of that region.

If you must choose a multi-storey hotel in a touristy area, try to choose one with its own solar energy units and purification plants, and, if you can bear it, no electricity-gobbling air conditioning.

10. *Adopt the local culture*. In other words 'When in Rome . . .'. This may seem obvious to some but if you are not aware of being in another country with different customs and habits is there any point in going there, other than for the sunshine? Enjoy the local cuisine, it will be invariably better than the locals' efforts at British food and if sensible precautions are taken, you will no more risk food poisoning than in a normal British restaurant. If you are in a hot climate, adopt the local habit of getting up early in the morning and taking a siesta after a midday meal. These habits have often evolved as a result of years of practice to find the most comfortable lifestyle for the climate.

11. *Get to know the locals*. The locals or host population of any destination make up the usual rich mixture of good, bad and colourful. When there are only a few tourists in any one area, contact with the locals is straightforward and often much hospitality is shown to tourists or 'guests' particularly if you can speak a few words of the local language. As numbers increase, so the relationship between guest and host becomes more of a financial transaction. You pay for a service such as a meal or room and the seller becomes interested in you mainly as a source of profit or income. With mass tourism in large centres it's obviously more difficult to overcome this; however, even in such areas only a certain proportion of the population is ever directly involved in the tourist industry. Many more are not, and live their own lives unrelated to the tourist world. So, how can you get an insight into these lives? True, many of them may not want to get to know you, but it's surprising how, despite the tourist hordes, the desire to be hospitable to guests still survives. The trick is being able to transform yourself from being simply a tourist to being a guest. Being yourself, you might say. There are a number of ways in which you can do this: you might make contact with local people or an organisation prior to leaving home who share a common hobby or interest with yourself, such as

sport, natural history, historical or cultural societies. Try visiting people in the same line of work as yourself to swop experiences, or visit something unusual such as a factory or school, where foreign visitors, even in tourist areas, may be unusual. Of course there's a limit as to how many can go to the local coal mine or hospital, but you'd be surprised how a letter from abroad prior to your arrival opens doors. Language barriers are overcome and as English speakers we are in a uniquely fortunate position, for it truly is the international language and there's generally someone, somewhere with a smattering of English to translate. Sign language and stick drawings get round a lot and it never stops good interaction with children, who are the best ambassadors of any country.

Think of yourself, how pleased and proud you'd be to show a foreigner round your town and work and you can see the appeal. A small token of appreciation in the form of something typically British always goes down well.

12. *Consider the impact of your visit on local resources*. Simply by being a tourist you are placing a strain on the capacity of the local infrastructure. In some cases even the innocent use of water can be the cause of tourist host resentment and deprivation, since in many poor countries tourists place a severe strain on the local water supply. The average tourist uses up to four times as much water as a local inhabitant and often water which is much needed by the local community is diverted to hotels to cope with tourist demand. There are many examples of this: at Djerba in Tunisia, 20 per cent of the water in the main supply network went to the large hotels, although 80 per cent of the dwellings in the town had no running water at all.[1] Again in Tunisia, in the Tozeur oasis, annual water consumption by tourists had increased from 500,000 litres in 1983 to 1.2 million litres in 1985. This amount of water would irrigate 124 acres of oasis land with 12,000 date palms. In Goa the demand for water to big hotels means that the villagers only get water for about one hour a day: in contrast the Taj Agoada Hotel gets 66,000 gallons per day; the Majorda Beach Resort gets 22,000 gallons; Didade de Goa Hotel 33,000 gallons and Bogmalo Hotel 44,000 gallons. With a little insight

and knowledge, you might help minimise unnecessary demand.
Other things to consider:

i) Whilst we're advocating that you 'go native', this need not include eating songbirds, frogs' legs, whale meat, turtle soup and the like. Often these dishes are only served in restaurants for the tourists out of sheer novelty value. But if enough tourists choose to ignore, or better still tell the management they definitely *don't want* these dishes, it could avoid a lot of unnecessary butchery.

ii) *Must* you demand floodlit tennis and golf and risky sports in remote places? If everyone does, it increases pressure on the holiday areas tremendously, in physical and resource terms.

iii) If you're going diving, trekking or viewing nature, even if you'd rather do it alone, it's better for the fauna and flora if you do it in small groups, less often. A continual stream of those who 'want to do it alone' is an expensive environmental luxury in areas under threat.

13. *Dress appropriately*. Find out about the cultural code of dress. Many of us, when on holiday, believe that we have the right to dress as casually as we want, no matter what the circumstances, without realising the offence we give. Just because we are not used to the extremes of heat and nobody knows us here does not make it any more right to enter a temple or mosque in shorts and a T-shirt than it would at home to go to church like this. If you want to dress like this stay in the hotel or by the swimming pool; don't go to the cultural sights or places of worship without being suitably clad.

In Islamic countries, women in particular need to be aware of what is considered locally acceptable. Buildings of worship and holy places such as temples, mosques, churches, cathedrals and shrines require care and tact.

14. *Enjoy local crafts and traditional arts*. This is a difficult one. There is a fine line between local arts and crafts developed for the tourist and the cheap replicas based on these arts and crafts which

have lost the true character and unique features of local traditions. Every country or region has something different to offer, and there is no doubt that the commercial interest of tourists has helped preserve and develop local traditions in a beneficial way. Nevertheless we should support these with discernment, encouraging only what is genuinely local. Find out more of the background of such crafts in terms of what they represent; also try to find out if the crafts are owned by the people who make them or if it's a middle man who profits by their sale. In many Third World countries cheap labour involving women and young children working long hours for appalling wages are used in the manufacture of souvenirs. Rich local businessmen buy these in bulk to sell on to street traders who also earn a pittance. Perpetuating this cycle is obviously not to be recommended. In India and the Far East the beautiful carpets for sale are often produced in sweatshop conditions.

15. *Souvenirs.* The whole question of souvenirs is obviously related to enjoying local crafts and traditional arts. The sale of holiday souvenirs is big business contributing a large amount to the overall economic turnover of tourism. Something like 12 per cent of all tourist spending goes on buying souvenirs or other goods.[2] Unfortunately many tourists are ignorant of the source of many of the souvenirs, however not only should we think of the cultural impact of the souvenir trade but also the impact of endangered flora and fauna. There are established conservation controls on the trade in rare wildlife set down by the Convention on International Trade in Endangered Species of Wild Fauna and Flora (CITES), and enforced by EC regulations. CITES has 103 signatory countries and reviews its controls every two years. It lists two categories of threatened species:

– endangered species (called Appendix I) which includes
sea turtles, snow leopards and almost all big spotted cats,
Asian and African elephants, rhinos and most crocodiles
and alligators. Trade in these and their products is
completely forbidden.
– vulnerable species (called Appendix II) which includes
some corals, most primates, cats, seal, whales and

dolphins and the majority of parrots. Trade in these is strictly limited and to bring them into the UK you must have a special permit from the Department of the Environment.

Common types of souvenirs whose purchase increases the risk of rare species becoming extinct are ivory trinkets, tortoiseshell bangles or earrings, reef corals or snakeskin purses. Birds' eggs, rare plants and mounted butterflies are also unnecessary victims of the trinket trade.

Unfortunately in many countries the control of trade in souvenirs made from endangered plants and animals is lax (witness the WWF plea to boycott tourism to Thailand in 1991 due to the Thais' repeated abuses of this legislation) and tourists are often offered goods made from these supposedly protected species. By buying them tourists help create a demand. By not buying them and, furthermore, by telling the shopkeepers *why* you are not, tourists can stop the supply and help conserve wildlife. Sometimes the issues are muddied, such as with the Cayman island turtles, where turtle conservation is linked with turtle 'farming' for desirable products, many of which are sold to tourists. It's best to avoid such souvenirs, even if they are fake. If you want to be certain, any wildlife product sold in a CITES country should be accompanied by appropriate documentation. The Department of the Environment issues two useful information leaflets outlining the regulations and species in danger; one leaflet is on endangered species in general and one is on endangered plants.

You are most likely to come across souvenirs of this type in Far Eastern countries such as Thailand, Indonesia, Hong Kong, Singapore and in Latin America, the Caribbean and increasingly Africa. Some countries in Europe such as Spain, Greece and Belgium act as distribution centres for goods made from the illegal trade in wild animal products. In particular the fur trade flourishes in Greece and sellers often give misleading advice to purchasers on what is permissible to bring back into the UK.

One particularly horrible aspect of the souvenir trade to beware of is the use of animals as props for souvenir photographs. The World Wide Fund for Nature has highlighted the plight of animals such as chimpanzees, lion and tiger cubs in Spain. These animals are imported illegally into Spain for use by unscrupulous photographers who entice tourists to pose with the animals for a holiday souvenir photograph. The animals are used when young and then discarded or killed when they get older and more troublesome to handle. In Third World countries tourists are often offered the opportunity to view a captured snake or animal in return for a small tip. While there has always been a use of animals for such purposes, the decline in wild habitat of many animals coupled with the pressures of increased tourist numbers has resulted in the numbers of many animals being decimated.

16. *Photography*. Photography as a leisure interest has boomed in the past decade with the advent of relatively cheap high quality cameras. It has much to commend it in helping people 'capture' their holidays forever and evoke memories of distant lands. Indeed one of the catchphrases for responsible tourists has been the exhortation by the Sierra Club (one of the most influential American Conservation Societies) to 'take nothing but photographs; leave nothing but footprints; kill nothing but time'. However, doing even those things out of context can be harmful: too many inappropriate photos and too many eroding footprints don't help! Care with 'souvenir' photography has been mentioned previously. There is another aspect to the subject – good manners and local customs in the country in question. In Third World countries in particular, many people are unnerved by photography and feel threatened by it and some object to being photographed by tourists and being regarded as something 'odd' to be photographed, how would *you* feel if some foreign person came up to you and, without a word, clicked away at you going about your business? (Indeed, the Hakka people in China, among others, believe that you steal a part of their soul if you photograph them, so do be aware and sensitive to these issues.) Payment is increasingly being sought by locals for being photographed and they can get threatening if no reward

is offered. It is, therefore, important to investigate the do's and don'ts of any particular country and exercise appropriate caution. Occasionally it is the failure to take photos that offends, as it may imply the dance or ceremony in your honour was not satisfactory; children, in particular, make keen models. There is no hard and fast rule.

Perhaps you might also consider that by concentrating on the photography of an occasion you may miss out on the spirit of the people or place you are visiting. While the camera, in the hands of a skilled photographer, can be capable of art, it can also be a barrier to communication. Leave it at home occasionally!

17. *Tipping, bargaining and giving money to beggars.* These are always awkward considerations for tourists caught between not wanting to be 'ripped off' by sharp practice and not appearing mean and small minded. The problem is most acute, as usual, in the Third World, where the gulf between wealthy tourists and poor locals is all too apparent. The situation, however, does also arise in developed countries. In the US, for example tipping is a way of life and is meant to be a reflection of the standard of service you receive. The best advice is to follow local custom: if locals tip then it's part of the local practice and you should too, though at the same rate. Unfortunately many employers in the tourist industry have got wise to this and often pay employees extremely low wages, safe in the knowledge that tips will make up the shortfall sufficiently to attract workers. You may feel, therefore, that tipping only encourages employers to pay low wages, but it is a fact of life in many countries and tourists will all be in the situation of 'do I tip, and if so, how much?' If you feel you have received good service then follow local custom. In some countries though, particularly in Asia, tipping is not traditional and the influx of tourists offering tips for everything can offend.

Bargaining or haggling is an area which has received more than its fair share of misinformation from tour operators and authors of holiday guides through crass generalisations. The following article in The *Ensenada News* gives one popular side of the equation:

> Mexico is known as a bargain hunter's paradise, from its economical hotels to its inexpensive gourmet dinners. But many tourists miss out on the greatest international sport since soccer – haggling! By paying the first price that is quoted when inquiring cost, you miss out on an age-old battle of wits and bluffing: that's the bargaining game.[3]

The other side is pointedly given in a leaflet produced by a group of Mexicans living in Oaxaca and Mexico City and distributed to tourists:

> Regarding the price of articles you may wish to buy, there is misinformation about so-called 'price haggling'. Tourist books have contributed to this mistaken concept. Almost always, the asked-for price is the correct one. In some cases, however, it is lowered to acquire food needed for the day. At such time the buyer acquires the article below cost without having the least idea of this. RULE: If you think it is too expensive, do not buy it.

Again the best answer is to try and find out local practice and follow it, although this is admittedly difficult for obvious tourists in Third World countries. A useful piece of information is to find out the average national wage and average wage of people in the region you are visiting; this gives a good indicator of what is value for money and what is not.

In many major cities and tourist resorts traders set prices which are unrealistic and aimed at gullible tourists. In more rural areas this is not the case. A good rule of thumb is to remember that traders in tourist resorts and cities who deal in tourist articles are generally very aware what things cost in developed countries and they are also used to bargaining and getting top prices. In remote areas, or when dealing with merchants who are not used to tourists, e.g. those selling fresh fruit or produce, more care is needed – the best bargain is when both parties are satisfied. This is perfectly possible in Third World countries because of the disparity of incomes.

Sometimes offering a higher price than is stipulated causes

offence. In Bali beautiful batik work and wood carvings are produced by local craftsmen for sale to tourists and some visitors feel that the prices are too low and think they are being generous in paying over the selling price. In fact the craftsmen work for the sake of their art rather than to make a profit. Their price is fixed to suit their purpose and the net result of many tourists being 'generous' is to cause prices to rise and the craftsmanship to deteriorate. A simple word of enthusiastic praise will go much further to express appreciation of the work.

Dealing with beggars and begging is something most of us find particularly difficult. Most developed countries have some form of social security which provides a safety net for those down on their luck. Accordingly, begging is an unusual phenomenon except in big cities, where requests for cigarettes or cups of coffee can be irritating. In Third World countries however there is very often no social security system and those individuals unfortunate enough to be seriously disabled, orphaned or homeless often have no option but to resort to begging. Once again, follow local practice. In India, for example, we witnessed more giving to beggars than anywhere else in the world. In countries such as these begging is accepted as part of society and people give on a varying scale, each according to his ability. Obviously you can't give to every beggar, but after finding out local practice, work out your own strategy, whether it's going to be small coins to several beggars, or a lump sum to one person you try to get to know. And stick to it. With children, giving pens or pencils is often greatly appreciated and this is a more sensible practice than giving money, for this is increasingly encouraging children to take up full-time begging at the expense of their education.

One constructive way to help is to donate money to a registered charity, either at home or in the country you are visiting, for a particular cause you are interested in. Sponsoring a local child through Plan International UK is an excellent and practical way to help (see Appendix).

18. *Reporting Back*. Take time if you used a travel agent or tour operator, to send in comments – not necessarily on the usual things like the standard of the food but on, for example, the helpfulness

of pre-trip briefings, observations of good/bad effects of tourism, suggestions to provide 'good tourism' practice.

A sense of balance

In summary, then, it is a sharpened sense of balance that is the essential ingredient needed for a good tourist philosophy. While the emphasis is strongly on local practice and customs, and the education of visitors, remember that tourists may often ironically find themselves in situations where local customs might, in fact, offend them and their values, both morally and from the standpoint of being a good tourist. It must always be remembered that tourists are guests in the country in which they are staying and as such they have a responsibility to act with tact and diplomacy, though not at the cost of limiting their freedom of choice. They still have the right to point out firmly but politely that they do not agree with a particular action or custom. Sufficient pressure will ultimately result in the questioning of an activity as regards its value or worth in today's world. Some examples of traditional activities which could provoke justified, but controlled, indignation are Spanish bullfights; the traditional capture and consumption of small song-birds in Southern Europe; the use of animals to entertain or transport tourists such as horse-drawn carriages; the hunting or fishing of wild, and sometimes rare, animals and fish.

It is difficult to set down hard and fast rules on this particular subject since reaction to these kinds of activities is largely dependent upon one's own point of view. Many might point out that the practice of hunting, fishing and shooting are the mainstays of much of the rural economy in Scotland and other parts of Europe and heavily influence the pattern of land use and the conservation of the countryside. Indeed many conservationists argue that these traditional practices are in fact beneficial to wildlife, helping to maintain a balance between predator and prey. So, as visitors we should seek to make informed decisions as to the impact of our behaviour and act accordingly, realising that there is not one thing *in isolation* that makes us good tourists, but an amalgam of many.

Chapter 7

Conservation Holidays

In Britain conservation holidays have been operating successfully since the 1950s and many of the leading organisations now form part of an international network of mainly charitable organisations providing these types of holidays. Each year many thousands of conservation volunteers spend their holidays and weekends making a practical contribution to the conservation of all aspects of the environment – from wildlife and the natural landscape to historical sites and man-made artefacts. As a holiday option a 'hands on' experience can provide an unforgettable holiday experience, an opportunity to care for the environment while gaining a deeper understanding of it. This chapter provides a comprehensive guide to the organisations offering conservation holidays so that you can choose the one most appropriate to your interests, needs and abilities.

Conservation holidays are available throughout the world and cover an enormous range of tasks. For example, bothy building in Glencoe; footpath maintenance in the Himalayas; archaeological digs in France; and Kangaroo surveys in the Australian outback. There is a project to suit all interests. The very nature of the work involved means that living conditions can be very basic and the work itself can be strenuous and uncomfortable. However, with a little preparation and a sense of humour these aspects will usually be a small price to pay for the experiences gained on the project. The active work on the holiday will not usually stop at the work site as projects are very much a team effort, with all volunteers expected to share in the domestic chores where appropriate. This is all part of the holiday as are the many activities that are organised for days/evenings off.

Conservation volunteers represent a workforce vital to charitable organisations and represent fieldwork that could not otherwise be achieved. Increasingly volunteers are also playing an important role in providing both financial and practical support to scientific research projects. For the volunteer the benefits are many, from the intense satisfaction of doing practical conservation work to the opportunity to work in beautiful and often remote surroundings. In addition to their educational value these holidays are usually inexpensive, giving value for money and representing a genuinely green holiday for the good tourist. It is not only the practical conservation achieved on these holidays that marks them out as green alternatives – other characteristics to consider include:

– conservation projects usually involve small groups of ten to twenty people. This ensures that fewer demands are made on the environment, groups are not considered to invade small local communities and while their practical contribution to the area is considerable, negative impacts are minimal.

– projects utilise small scale local accommodation and by doing so support the local community.

– many conservation projects support research important to the survival of local ecosystems.

– projects often involve working with and for the benefit of the local community.

– economic contribution to the local economy is often high, as projects buy their provisions locally.

– many projects are directly beneficial to local communities, some providing housing and employment following their conclusion, e.g. farm restorations.

The sense of achievement and the skills acquired on conservation projects can become addictive and many opportunities exist for involvement and interest to continue long after the holiday is over. For example the British Conservation Trust for Volunteers and Scottish Conservation Projects organise comprehensive training courses teaching conservation skills. Many organisations such as the National Trust have active local groups throughout the country. Membership of conservation organisations can continue your holiday interests. Brochures available from the contact addresses below will provide further details and all the necessary information, including details of living conditions and levels of fitness necessary for each project. Because of the work involved in these holidays you need to be well prepared – organisations will advise as to suitable provisions, and insurance and a tetanus injection are both strongly advised.

Conservation holidays available in the UK

THE BRITISH TRUST FOR CONSERVATION VOLUNTEERS.
BTCV's conservation working holidays offer hundreds of opportunities to take a break tailored to act in harmony with the natural world. Nearly 600 Natural Break week-long and weekend holidays are organised in some of the most beautiful parts of England, Wales and Northern Ireland. Designed to give you the chance to protect endangered wildlife and repair characteristic landscape features, these holidays run every day of the year and are perfect for every 'good tourist' over the age of sixteen.

No previous experience is necessary as full training is given in all traditional conservation skills from restoring eroded earthworks on an archaeological site on the Isles of Scilly, to building the first North Pennines nature trail at Allenheads, the highest village in England.

The average price for a Natural Break is £29.50 including all food and accommodation, which can vary from a village hall to a field studies centre, and time off is given for exploring the surrounding landscapes.

BTCV's International Working Holidays offer over thirty opportunities to join local European communities on a range of cultural and ecological conservation projects lasting between one and three weeks. Holidays range from building a tortoise village in Corsica to creating a bird reserve in the grounds of a Belgian château, giving the unique chance to try traditional conservation projects in superb European environments.

Prices for the International Working Holidays vary from £25 including food and camping accommodation in Holland, to £250 including food, accommodation and flight to Romania. Details of BTCV's Natural Breaks and International Working Holidays are available from: BTCV, Room GT, 36 St Mary's Street, Wallingford, Oxon, OX10 0ED. Tel. 0491 39766 (24hr) Fax. 0491 39646.

CATHEDRAL CAMPS.

Cathedral Camps aim to provide willing hands for the maintenance and restoration of Britain's rich heritage of cathedrals, abbeys and other significant Christian buildings. Work varies from routine cleaning to stonework restoration and is supervised by experienced craftsmen/professional conservators. Camps are a week long, from mid-July to September and cost approximately £35 including food and basic accommodation. Volunteers work a thirty-six-hour week with a day and a half and all evenings free. Age range is sixteen to thirty and first-timers need a letter of recommendation. Volunteers receive an admission card for one year for all English cathedrals.

Further information from: Manor House, High Birstwith, Harrogate, North Yorks, HG3 2LG. Tel. 0423 770385.

FESTINIOG RAILWAY COMPANY.

The company operates and maintains a 150-year-old narrow-gauge railway between Porthmadog and Blaenau Festiniog and relies upon volunteers and enthusiasts to keep the line open. Volunteer projects are varied and include park and garden work, line and locomotive maintenance, staffing shops, cafes and booking offices. No experience is necessary and training is provided. Age range is sixteen and over, and breaks are for one week or more, all year round. Volunteers must be fit, and organise their own accommodation and transport.

Further information from: Volunteer Officer, Harbour Station, Porthmadog, Gwynedd, LL49 9NF. Tel. 0766 512340.

GROUNDWORK (IRELAND).

In association with the Irish Wildlife Federation, Groundwork carries out important conservation work in some of Ireland's most inaccessible areas. Work is strenuous and conditions can be uncomfortable. Age range is sixteen and over, breaks for one or more weeks, June to August. Self-catering accommodation and insurance is included in the IR£10 booking fee (IR£5 for successive weeks).

Further information from: Groundwork (Ireland), 43 Bayview Drive, Killeney, Co Dublin.

NATIONAL TRUST ACORN PROJECTS.

Formed at the end of the nineteenth century, the Trust is a charity concerned with the preservation of places of historic interest and natural beauty including houses, moors, mountains and coastlines. Over 200 Acorn projects are organised each year from March to October at many of the Trusts' 300 properties throughout England, Wales and Northern Ireland. Projects include archaeological digs, footpath restoration and erosion control. Ages range from seventeen and over and volunteers work an eight-hour day with a half-day and all evenings free, for one or more weeks. Accommodation varies from village halls to National Trust cottages. One week costs approx. £30 including food, accommodation and insurance. Volunteers completing a project receive one years' free admission to NT properties. The Trust also has a network of active local groups. Application forms are available in January and early booking is advisable.

Further information from: NT Acorn Projects, Volunteer Unit, PO Box 12, Westbury, Wiltshire, BA13 4NA. Tel. Westbury (0373) 826826.

NATIONAL TRUST FOR SCOTLAND THISTLE CAMPS.

The NTS cares for over 100 properties in Scotland including castles, battlefields, islands and gardens. Approximately twenty-five conservation camps are held each year between March and October. Projects include bothy construction, croft work, habitat management and footpath maintenance. Age range is sixteen to seventy, and camps last for one or more weeks. Volunteers work an eight-hour-day with one day and all evenings free. Accommodation varies as for NT Acorn Camps. One week costs £18 including food, accommodation and insurance (£9 for UB40 holders). Volunteers completing forty hours conservation work receive a Youth in Trust sweatshirt. NTS also has a number of active local groups throughout Scotland. Information is available in January and early booking is advised, especially for island projects.

Further information from: Jim Ramsay, NTS, 5 Charlotte Square, Edinburgh, EH2 4DU. Tel. 031 226 5922.

THE ROYAL SOCIETY FOR THE PROTECTION OF BIRDS.

In addition to its excellent campaigns and education programmes, the RSPB also has an extensive network of volunteers helping in its work. Voluntary work includes national research and survey projects, recording birdlife in Britain throughout the year and acting as wardens on specific sites. Perhaps the best known of these is Operation Osprey in Speyside. Volunteer wardens keep a twenty-four-hour watch over the Osprey, recording their activities and acting as guides to the many visitors to the site. Age range is eighteen and over and breaks are for one week or more, March to August. Volunteer cooks are also required and will have some or all travelling expenses paid depending on the length of stay. Full-board and camping accommodation is provided at a nearby farm for around £25 per week.

Further information from: RSPB, Reserves Management Dept, The Lodge, Sandy, Bedfordshire, SG19 2DL. Tel. 0767 680551.

SCOTTISH CONSERVATION PROJECTS.

This charity, formed in 1984, exists to promote the practical involvement of people in conserving the scenic and wildlife heritage of Scotland and is the Scottish arm of BTCV. Seventy Action Breaks per year represent an extensive range of conservation work throughout Scotland and the Isles. Projects include pond clearance, wildlife management, drystone dyking and bridge restoration. Breaks are usually ten days with one day and all evenings free. Age range is sixteen to seventy. Accommodation varies from a mat on a bothy floor to comfortable chalet houses. Costs begin at around £20 and include food, accommodation and insurance. Volunteers must be members of SCP or BTCV to join an Action Break. SCP also has an extensive local group network and runs special campaigns such as Operation Brightwater, a three-year initiative linked to the threats to Scotland's freshwater lochs, rivers and coastline.

Further information from: SCP, Balallan House, 24 Allan Park, Stirling, FK8 2QG. Tel. 0786 79697.

SCOTTISH FIELD STUDIES ASSOCIATION.

The Field Centre at Kindrogan was established in 1963 to foster a greater public awareness and understanding of the Scottish countryside. It runs courses for school and university students on all aspects of the environment with additional courses for adults on general countryside topics. There are weekend or week-long field courses. An approximate weekly charge for a course is £177.

Further information from: Kindrogan Field Centre, Enochdhu, Blairgowrie, Perthshire, PH10 7PG Tel. 025 081 286.

THE WATERWAY RECOVERY GROUP.

Despite their importance as wildlife habitats and sources of recreation and potential transport systems, many of Britain's extensive canal networks are sadly neglected. The Waterway Recovery Group co-ordinates the many local societies/trusts who are working to correct such neglect and volunteers work on restoration projects to restore the canals to their former glory. Work may include dredging, bricklaying and excavation work on canals throughout the UK. Full training is given with experts on site. Age range is sixteen and over and breaks last one week, from March to October and during the Christmas season. Accommodation is basic and full board costs approximately £25 per week.

Further information from: The Waterway Recovery Group, Neil Edwards, Canal Camps, 249 Avenue Road, Witham, Essex, CM8 2DT. Office HQ, Tel. 0926 511634.

Conservation holidays available – Europe and beyond

AMERICAN HIKING SOCIETY, VOLUNTEER VACATIONS PROGRAMME.

The Society maintains hiking trails throughout America, using enthusiastic volunteers for a variety of tasks including research, observation, guiding and trail maintenance. Volunteers must be physically fit and a driving licence is useful. Accommodation is usually provided and due to the isolated location of many of the work sites this is often camping accommodation. Age range is sixteen and over with breaks of two or more weeks, all year. These posts are very popular and an early application in November/December is advised. Volunteers must arrange their own transport. For $3 AHS will provide a directory of these and 1,000 similar voluntary posts in America.

Further information from: The Director, VVC, AHS, 1015 31st Street, N.W. Washington D.C. 20007, U.S.A. Tel. 0101 703 385 3252.

ASSOCIATION OCCITANE POUR LA DEFENSE DE LA FÔRET (ASSODEF).

An organisation concerned with the conservation of forests and the education of people as to their importance and value. Projects include some of the more unusual aspects of forest care such as providing observation posts and a solar powered information centre. Age range is eighteen and over with breaks lasting two or more weeks, July to September. Afternoons are free to enjoy the many organised leisure activities. Food, accommodation and insurance are provided and the registration fee is FF500.

Further information from: ASSODEF, Hôtel de Ville, Chemin des Loutabas, 13860 Peyrolles, France. Tel. 010 33 425 78005.

AUSTRALIAN TRUST FOR CONSERVATION VOLUNTEERS.

ATCV promotes management and care of the environment through practical conservation projects throughout Australia. Projects include urban research in addition to the more usual conservation work in Australia's National Parks and established walking areas. Minimum age is seventeen and projects cost approximately AU$500 for six weeks, including food, accommodation and travel while on the project.

Further information from: ATCV, National Director, PO Box 423, Ballarat 3350, Victoria, Australia. Tel. 010 61 53327490

BTCV INTERNATIONAL CONSERVATION WORKING HOLIDAYS.

Following thirty years' success in Britain, as part of a major new initiative, the Trust is now expanding its operations into Europe. The aim is to introduce the volunteering ethic to communities abroad and to encourage greater communication and understanding between different communities. The International Working Holidays operates along similar lines to the BTCV's Natural Breaks, with projects throughout Europe including bridge restoration in Portugal and pond management in Luxembourg. Prices start at around £100 per week including travel, food, basic accommodation, insurance and transport from a pick-up point. Project leaders are bilingual and volunteers should obtain an E111 form from a social security office prior to departure.

Further information from: BTCV, 36 St Mary's Street, Wallingford, Oxfordshire, OX10 0EU. Tel. 0491 39766.

THE CHRISTIAN MOVEMENT FOR PEACE.

Volunteers work on international conservation projects worldwide. Recent projects include wildlife habitat management in Germany and organic farming in America. Age range is eighteen and over and breaks are for two to three weeks, usually during the summer months. There is a registration fee of £24 and volunteers must pay their own travel expenses to the site where food, accommodation and insurance is provided.

Further information from: Christian Movement for Peace, Bethnal Green United Reformed Church, Pott Street, London, E2 0EF. Tel. 071 729 1877.

CORAL CAY CONSERVATION.

Qualified divers needed to help survey a proposed marine park in Belize's coral reef which is under threat from pollution and mass tourism. One month projects from April–Dec. costing around £1,500 all-in.

Further information from: Sutton Business Centre, Restmor Way, Wallington, Surrey, SM6 7AH. Tel. 081 669 0011 Fax. 081 773 0406.

Conservation Holidays

EARTHWATCH.

Founded in 1971, Earthwatch is an international charitable organisation providing monetary and practical support for scientific research projects worldwide, one hundred and twenty projects in forty three countries. Described as a merchant bank for the field sciences, Earthwatch aims to match a volunteer's experience and interest to the needs of field scientists. Through such practical involvement Earthwatch hopes to improve human understanding of the processes affecting our planet and the diversity of its life forms. Volunteers willing to spend their holidays working for the environment share the project costs and work alongside world experts, studying anything from the social life of the Kangaroo to birdlife in the Mediterranean. Accommodation, conditions and cost vary according to the project and a basic membership fee of £22 is payable. Volunteers must be over sixteen years and projects run year-round, usually lasting two to three weeks.

Further information from: Sally Moyes, Earthwatch Europe, Belsyre Court, 57 Woodstock Road, Oxford, OX2 6HU. Tel. 0865 311600.

EUROPE CONSERVATION.

This is a new environmental organisation which aims to increase Europeans' knowledge about European natural resources through holidays and research programmes in parks and natural reserves. The focus is upon practical education and all programmes include preparatory lessons. Projects include research expeditions for bear monitoring in Spain and archaeological camps in Sardinia and Turkey. There are also educational programmes based at nature centres throughout Italy where family groups are welcome. Ecological research is the main activity of the group. Volunteers must be members of Europe Conservation to join a project and this costs £10 per year. Age range is eighteen and over. Costs include food and accommodation and vary widely depending on the location and duration of the project.

Further information from: Europe Conservation, via Fusetti, 14–20143 Milan, Italy. Tel. 010 39 2 5810 3135 Fax. 010 392 8940 0649.

GENESIS 11 – CLOUD FOREST.

The Talamanca cloud forest is being preserved by private owners for research and recreational activities. Volunteers are needed to help with conservation tasks in the forest, including trail maintenance, reforestation and fencing. Training is provided and physical fitness is particularly important for these holidays due to the terrain and high altitude (2,360 metres). Age range is seventeen and over with holidays of six weeks or more duration, all year. Volunteers work a twenty-five-hour week with weekends free, so there is plenty of time to explore the area. Volunteers contribute $50 per week for dormitory accommodation, meals and laundry, but must make their own travel and insurance arrangements.

Further information from: Genesis 11 – Talamanca Cloud Forest, Apartado 655, 7050 Cartago, Costa Rica, Central America. Tel. 010 506 795243.

INTERNATIONAL VOLUNTARY SERVICE,
UNITED NATIONS ASSOCIATION INTERNATIONAL YOUTH SERVICE.

These organisations co-ordinate international conservation workcamps worldwide. Many of the camps include a study element and families are accepted on some, e.g. in Denmark and Finland. Characteristics of the camps vary but as a rough guide the minimum age is eighteen, most camps run from June to September and volunteers work a thirty-hour-week for two weeks or more. Food, accommodation and insurance is provided and some workcamp experience is preferred. A membership and registration fee is payable to use the service. When applying, specify which country you are interested in and specify conservation work as your main interest as both organisations also operate community and other schemes.

Further information from: IVS, 162 Upper New Walk, Leicester, LE1 7QA. Tel. 0533 549430. UNA International Youth Service, Welsh Centre for International Affairs, Temple of Peace, Cathays Park, Cardiff, CF1 3AP. Tel. Cardiff 223088.

LA SABRANENQUE CENTRE INTERNATIONAL.

Founded in 1969, this charitable organisation aims to restore and reconstruct abandoned rural farms and villages so that they may be used once again. Their first project was the restoration of a small village outside Avignon, southern France. Following the successful completion of that project, La Sabranenque is now working on other sites in the area and supervises similar projects in Gnallo, northern Italy and Ibort in Spain. Project tasks may include masonry, tiling, paving and tree-planting. Age range is eighteen and over and breaks last two or more weeks, June to August. There is at least one day off with organised activities available. These are international work camps so don't worry too much if you don't know the local language. Projects cost FF65 per day including food and accommodation and there is a registration fee of FF120.

Further information from: La Sabranenque Centre International, 30290 Saint Victor la Coste, France. Tel. 010 33/66500505.

Rural redevelopment is a popular conservation theme throughout France and there are many organisations doing similar work to La Sabranenque. Try one of these if the Sabranenque camps are full, or you are interested in another area: Etudes et Chantiers International, 33 Rue Campagne Premiere, 75014, Paris; Jeunesse et Reconstruction, 10 Rue de Trevise, 75009 Paris; Pro-Peyresq Secretariat, c/o Windberg 290, 1810 Wemmel, Belgium; Rempart, 1 Rue des Guillemites, 75004 Paris.

UNIVERSITY RESEARCH EXPEDITIONS PROGRAMME.

Based in California, UREP enables volunteers to participate in scientific research projects worldwide. Recent research topics include insect and flora relationships in Costa Rica, Polynesian rock carvings in Hawaii, and deer tracking in California. Age range is sixteen and over and volunteers must be fit. Breaks last for two or more weeks from Dec–Feb and June–Oct. Volunteers pay an equal contribution towards the cost of the project (around $1,000) which includes equipment, camping accommodation, meals and ground transportation. Volunteers organise their own transport to the research site. Applications must be in two months before the trip begins.

Further information from: The Secretary, UREP, Desk L10, University of California, Berkeley, CA 94720, USA. Tel. 0101 415 642 6586.

Conservation Training Courses

FIELD STUDIES COUNCIL.

This excellent organisation has nine residential field centres throughout southern England and Wales, offering a wide variety of courses to promote 'environmental understanding for all'. Most of the courses last a week and cover a diverse range of topics from ecology and conservation to history and architecture; landscape and climate to birds and other animals. Course participants must be members of FSC to attend courses and this costs £12 per year if you receive journals, £5 per year if not. Inclusive prices vary from £80 for a weekend on spiders to £190 for a beginners guide to marine biology for divers. Family courses are available.

Since 1978, FSC has operated an extensive overseas expedition course programme. Each overseas course is led by experts in the field and they cover topics as diverse as New Year in the Everglades and flowers of the Pyrenees. Prices are all inclusive and start at £250 for a weekend studying birds in Holland.

Further information from: FSC, Central Services, Preston Montford, Montford Bridge, Shrewsbury, SY4 1HW. Tel. 0743 850674.

OPERATION RALEIGH

'The ultimate expeditions to develop leadership skills.' Challenging expeditions assisting local communities, contributing to world conservation and undertaking scientific research. Must be selected to take part by undergoing various tests. Average price from £4,900 + VAT for 4 weeks in southern Chile.

Further information from: The Powerhouse, Alpha Place, Flood Street, London SW3 5SZ. Tel. 071 351 7541 Fax. 071 351 9372.

SCOTTISH CONSERVATION PROJECTS.

In addition to their wide-ranging conservation holiday programme, SCP also has an extensive conservation training calendar supported by the Countryside Commission for Scotland and the Nature Conservancy Council. The courses run mainly in autumn/winter and generally last two to three days over a weekend. Prices start at around £21 for SCP/BTCV members and include food, accommodation, insurance and transport from the nearest railway station. Courses range from basic to advanced level and cover many aspects of conservation work including dyking, footpath management and qualifications in chainsaw use. Age range is sixteen and over.

Further information from: SCP, Balallan House, 24 Allan Park, Stirling, FK8 2QG. Tel. 0786 79697.

SUNSEED DESERT TECHNOLOGY

Working to contain or reclaim deserts and running a research and development centre near Almeria in southern Spain and a re-afforestation programme in France, this charity offers voluntary work holidays. Various themes for study: use of sun and wind; pottery; role of trees. Approx. cost per week £45.

Further information from: PO Box 3000, Timworth, Bury St Edmunds, IP31 1HP. Tel. 0284 728863 Fax. 0284 728240.

NB: When applying to projects overseas remember to enclose an international reply coupon. For projects in the UK, enclose a stamped addressed envelope.

Chapter 8

Voluntary Work Holidays

Voluntary work can be anything from three years' teaching in Africa as part of the Voluntary Services Overseas organisation, to spending three hours a week as a volunteer in your local Oxfam shop. While representing two extremes of a diverse range of work being done around the world, these two examples have a number of important characteristics in common which go some way towards defining voluntary work:

– in each instance, a volunteer is giving something of themselves to help others – their time, effort, interest and support. For their help the volunteer may receive expenses or pocket money, but most importantly he/she gains the satisfaction of helping others and learning from the experience. The key word in voluntary work is 'involvement'.

– the organisation for which the volunteer is working is able, through willing hands, to achieve tasks not otherwise possible.

– somewhere, a third party is benefiting from the help offered, whether it be a holiday made possible by someone volunteering to push a wheelchair, or an organic farmer able to continue his business.

A clear and universal definition of voluntary work is difficult to state due to the diversity of opportunities, organisations and work. Generally, in return for work, volunteers receive food, accommodation and occasionally travelling expenses and a small amount of pocket money. Opportunities exist worldwide and there are projects to suit all interests and capabilities.

As a holiday, voluntary work represents an interesting alternative for the good tourist, many of their 'green' features being similar to those of conservation holidays. Voluntary holidays are often inexpensive and though the work can be strenuous and living

Voluntary Work Holidays 133

conditions basic, the community spirit achieved guarantees value for money.

As for conservation holidays, volunteers need to be well-prepared; insurance cover and a tetanus injection are both advised. Volunteers should be very sure that they know what to expect on the trip and what is expected of them: voluntary work is *not* simply a cheap holiday. Voluntary holidays are *working* holidays, so if the information says, twenty-four-hour care, then that is what is required of volunteers. If possible, it can be useful to allow yourself a few days non-working holiday in which to recover before returning to work!

The voluntary-work holiday suggestions given below are somewhere between the two extremes given earlier and assume that most people have one to four weeks holiday at a time to play with.

The UK has three main organisations which run workcamps and recruit volunteers for international voluntary workcamps overseas. These are outlined first and represent a useful information source and offer a huge choice of voluntary work to choose from. Following these is a listing of smaller organisations operating in Britain and abroad. These are listed under five categories: archaeology; social and community work; religious centres; children and youth care; and animals and agriculture. Finally, if you want to continue your voluntary work commitment after a voluntary working holiday then there are many opportunities to do so and some national information addresses are provided at the end of the chapter. For local agencies Yellow Pages or a Citizens Advice Bureau may be able to help.

International Co-ordinating Bodies in Britain

CHRISTIAN MOVEMENT FOR PEACE.
CMP is an international organisation working for peace and justice, offering voluntary services in areas of need and promoting community self-help. Voluntary projects operate throughout Europe, America, Canada and North Africa. Recent projects include play schemes in Northern Ireland, tool production in

Holland and building a sewage plant in West Germany. CMP also runs special theme projects e.g. building a float for the Notting Hill carnival to highlight the problems in Namibia and supporting the campaign for the homeless in London. Age range is eighteen and over with stays of two to six weeks during the summer. Food, basic accommodation and insurance is provided for volunteers, and there is a registration fee of £15 for UK camps and £24 for overseas to join CMP.

Further information from: CMP, Bethnal Green United Reform Church, Pott Street, London, E2 0EF. Tel. 071 729 7985.

INTERNATIONAL VOLUNTARY SERVICE.

IVS is the British arm of Service Civil International. It was founded in 1920 to promote peace, international cooperation and friendship through voluntary work. SCI has branches in twenty-five countries organising over 350 short-term voluntary projects each year. Projects vary according to the needs of the individual country and recent projects have included play schemes in Dublin; making radio programmes in Illinois and working with disabled people in Barcelona. Age range is eighteen and over and volunteers work an unpaid forty-hour week for two weeks or more. Meals and basic accommodation are provided and volunteers are expected to take an active part in the team life of the project. There is a membership fee of £25 (£15 for students and £10 for the unwaged) after which volunteers pay a registration fee of £20 for UK workcamps and £40 for overseas projects.

Further information from: IVS, 162 Upper New Walk, Leicester, LE1 7QA. Tel. 0533 549430

UNITED NATIONS ASSOCIATION, INTERNATIONAL YOUTH SERVICE.

Based in Wales, this IYS co-ordinates short-term international voluntary projects for British applicants and aims to encourage international understanding and promote the development of community projects. Voluntary workcamps are held throughout Britain, Europe, America, Canada, India and Africa, with recent work projects including bricklaying in Rome, constructing a museum in Poland and organic farming in Denmark. Age range is eighteen and over with stays lasting two or more weeks. Volunteers are unpaid but meals and basic accommodation are provided. There is a registration fee of £25 for UK camps or £35 for overseas projects.

Further information from: UNA, IYS, Welsh Centre for International Affairs, Temple of Peace, Cathays Park, Cardiff, CF1 3AP. Tel. 0222 223088.

Archaeology

There is a huge range of archaeological digs and research projects available worldwide, but they are particularly extensive in Britain, France, America and Israel. Volunteers may be involved in the actual excavation work or in the running of the camp. While experience is not essential for many of the volunteer posts, it is advisable to gain some experience on a British dig prior to going overseas. The organisations cited below represent the main sources of information available.

ARCHAEOLOGY ABROAD.

This organisation provides a similar service to the Council for British Archaeology, publishing an annual information bulletin. The bulletin is available in March, with two updates in spring and autumn. Enquiries about digs in specific countries are also welcome.

Further information from: The Secretary, Archaeology Abroad, 31–34 Gordon Square, London, WC1H OPY.

ARCHAEOLOGICAL INSTITUTE OF AMERICA.

The Institute publishes an annual bulletin listing archaeological sites throughout the US where volunteers are needed. Detailed information is given on board and lodging, ages and experience necessary, training opportunities and costs. The bulletin is available in January and costs $10.50 including postage.

Further information from: Archaeological Institute of America, 675 Commonwealth Avenue, Department GG, Boston, Massachusetts 02215. Tel. 0101 617 353 9361

THE COUNCIL FOR BRITISH ARCHAEOLOGY.

In addition to organising digs throughout Britain, the council publishes an informative bi-monthly newsletter *British Archaeological News*. This lists sites where volunteers are needed, giving details of accommodation, location, expenses etc. Conditions vary depending on the nature of the dig, but basic accommodation and a small allowance is usually provided. Volunteers work two or more weeks and need to be physically fit. Projects are available all year round and families are often welcome. Annual subscription for the newsletter is £8.50.

Further information from: The Council For British Archaeology, 112 Kennington Road, London, SE11 6RE. Tel. 071 582 0494.

Social and Community Projects

THE ACROSS TRUST.

A registered charity, the Trust organises ten-day holidays and Lourdes pilgrimages for the sick and disabled. Volunteers include nurses, doctors, chaplains and general helpers who care for the group twenty-four-hours a day, living together as a 'family' throughout the holiday. Volunteers must be fit and dedicated to the work necessary on the project. The holidays run from Easter to November each year, exact dates given on application. As with the duties on the holidays, the cost is divided between all members of the party, approximately £300 all inclusive.

Further information from: The Across Trust, Crown House, Morden, Surrey, SM4 5EW. Tel. 081 540 3897.

ASSOCIATION des PARALYSES de FRANCE.

Fifteen hundred volunteers help the Association in providing holidays for the handicapped and paralysed each year in July and August. The nature of the work requires volunteers to be fit and enthusiastic with a sense of humour. You need to be eighteen or over and be able to work for four weeks or more. Expenses and a small amount of pocket money is paid by the Association to volunteers completing their stay.

Voluntary Work Holidays 137

Further information from: The Holiday Officer, A.P.F. 17 Boulevard Blanqui, Paris 75013. Tel. 010 3314 580 8240.

COUNCIL ON INTERNATIONAL EDUCATIONAL EXCHANGE.

A non-profit making organisation sponsoring voluntary service projects for young people in the USA and abroad. Issues such as hunger, literacy, unemployment and development are explored while working with volunteers from other countries. Three-week projects as a rule, and many worldwide projects to choose from. There is an application fee of $100 plus transportation. Room and board provided at workcamp.

Further information from: 205 East 42nd St, New York, NY 10017 USA. Tel. 0101 212 661 1414.

NIGERIA VOLUNTARY SERVICE ASSOCIATION.

NIVOSA organises international workcamps throughout the states of Nigeria to promote international understanding and community cooperation. Volunteers help in all aspects of building and construction work, providing schools, hospitals and other community amenities for villages. Age range is eighteen and over and workcamps are of two weeks duration July to September. Volunteers work a six-hour-day and events are organised for free time. Most of the workcamps are preceded by an orientation/leadership course which all volunteers must attend. Volunteers pay own travel expenses.

Further information from: General Secretary, NIVOSA, GPO Box 11837, Ibadan, Nigeria.

PASCHIM BANGA SAMAJ SEVA SAMITY.

A voluntary organisation based in Calcutta, Banga recruits over 300 volunteers each year to help maintain a variety of social institutions throughout India, including hospitals, schools, and libraries. Volunteers must be fit as the work may be strenuous. Age range is sixteen to thirty, and volunteers work for three or more weeks in Jan/Feb, June/July or October to December. Food, accommodation and a small amount of pocket money is provided. Volunteers pay their own travel expenses.

Further information from: P.B.S.S.S, 191 Chittanjan Avenue, Calcutta – 700 007, India. Tel. 010 9133 39 7631.

THE SWALLOWS IN DENMARK.

The Swallows raise funds for aid organisations in India and Bangladesh by collecting, recycling and selling second-hand furniture and clothes. International workcamps are held during the summer when volunteers help at the recycling camp. Volunteers work an eight-hour-day, a six-day-week with full board and accommodation provided, but volunteers pay their own travel expenses. Families are welcome and camps are of one to three weeks duration.

Further information from: The Co-ordinator, U-Landsforeningen Svalerne, Osterbrogade 49 2100 Copenhagen O, Denmark. Tel. 010 45 31261747.

A similar project runs in Finland as part of the International Emmaus movement. You need to be eighteen or over and projects run for two or more weeks during the summer.

Further information from: Emmaus Westervik, 10600 Ekenås/Tammisaari, Finland. Tel. 010 35 811 25440

3 H'S FUND.

The fund aims to provide large group holidays for physically handicapped people and needs able-bodied volunteers to help ensure that the holidays run smoothly and that a good time is had by all. Most of the holidays are in the summer months and may include trips abroad. Volunteers work for one week or more and must be eighteen or over. Full-board, accommodation, insurance and travel is provided for UK holidays, with a small contribution payable by the volunteer for European holidays.

Further information from: 3 H's Fund, Holiday Organiser, 134 London Road, Southborough, Tunbridge Wells, Kent, TN4 0PL. Tel. 0892 511928.

Religious Centres

THE IONA COMMUNITY.

An ecumenical centre on a small island off the west coast of Scotland, the Iona community receives groups of people and many day visitors each season from March to October. Volunteers, aged eighteen and over, live as part of the community and perform a variety of work including housekeeping, guiding, maintenance and catering. Volunteers work for six weeks, six days a week, and are expected to share in the life of the community, including worship and social activities. Full-board and shared accommodation is provided and help with travel expenses is available for travel from within the British Isles.

Further information from: The Iona Community, Iona Abbey, Isle of Iona, Argyll, Strathclyde, PA67 6SN. Tel. 06817 404.

VALAMO MONASTERY.

A religious community since the twelfth century, Valamo is a popular ecumenical pilgrimage site. Volunteer tasks involve all aspects of community living including gardening, catering and generally being involved in the day to day life of the Monastery. A thirty-four-hour week allows plenty of time for relaxation, including excursions, discussions, saunas, forest walks, worship or lectures at the local academy. Age range is eighteen and over. Most volunteers work for two weeks in July and August but there are also year round projects. Full board and hostel accommodation is provided, but volunteers must make their own travel arrangements.

Further information from: Brother Chariton, Valamo Monastery, SF – 79850, Uusi-Valamo, Finland. Tel. 010 35 72 619 11.

Youth and Child Care

BIRMINGHAM YOUTH VOLUNTEERS.

BYV provide holidays in Wales during the summer holidays for under-privileged children. Volunteers are needed to help with the twenty-four-hour care of the children, including supervision and taking part in group activities. Minibus drivers are required, aged twenty-one and over with a clean driving licence and cooks who cater for up to sixty people. Stays are for one week only, July to September and volunteers must be seventeen and over. Meals and accommodation are provided plus travel costs from Birmingham to the holiday location.

Further information from: Co-ordinator, BYV, 3rd Floor, 24 Albert Street, Birmingham, B4 7UD. Tel. 021 643 8297.

FLYSHEET CAMPS SCOTLAND.
A registered charity, Flysheet has been running its self-sufficient children's farm and wilderness centre since 1980. It aims to enable children to experience living at close quarters with nature and gain experiences completely different to their normal way of life. Volunteers are involved in all aspects of the camps including caring for the children, organic farming, maintaining rare breed stocks and catering. The camp is very isolated and has no electricity so volunteers must be adaptable and have a sense of fun! Age range is eighteen and over and stays are for two or more weeks, June to October. Basic accommodation is provided and volunteers share food costs, approximately £2 per day. Volunteers must make their own way to the farm.

Further information from: The Resident Organiser, FCS, Finniegill Children's Farm, Lockerbie. Tel. 05766 211.

LIVERPOOL CHILDREN'S HOLIDAY ORGANISATION.
LCHO provides residential holidays in the British countryside for children who would not normally have a summer holiday away from their home. Volunteers are required for the twenty-four-hour care of the children, including supervision, activities, catering etc. Volunteers must be eighteen or over, physically fit and stay for one week or more during the summer holidays. Meals and accommodation are provided and expenses paid, including £30 per week pocket money. Inexperienced volunteers must attend a week long residential training course prior to their holiday.

Further information from: LCHO, Room L8, Wellington Road School, Wellington Road, Liverpool, L8 4TY. Tel. 051 727 7330.

Similar schemes to LCHO operate throughout Britain. Contact: London Children's Camp, 33 Farren Road, Forest Hill, London, SE23 2DZ; or Sunshine Holiday Home, Allonby, Cumbria, CA15 6QH.

NANSEN INTERNASJONALE CENTER.
Twenty-five kilometres from Oslo, the Brievold activity and relief centre aims to help children and young people with special needs. During the summer months up to twenty children stay at the centre each week, where sports, hobbies and the care of animals are used to motivate the children in all areas of their life. Volunteers help in all aspects of the centre's work and twenty-four-hour involvement is necessary. They must be aged twenty-two and over, and work for two weeks or more, June to August (or all year). Volunteers must be physically fit and previous experience is advantageous. Full-board, accommodation and NKr300 per week is provided, but volunteers must make their own travel arrangements.

Further information from: The Director, NIC, Barnegarden, Breivold, Nesset, 1400 Ski, Norway. Tel. 010 472 94 67 15.

Animals and Agriculture

BRITAIN–CUBA RESOURCE CENTRE.
The centre offers an unusual month-long package of voluntary work and guided tours providing an insight into Cuban life since the 1959 revolution. The voluntary projects involve agricultural or construction work and are strenuous, so volunteers must be fit. They need to be seventeen or over, they work a four-and-a-half-day, thirty-five-hour week and the other two days each week are spent visiting schools, hospitals, factories etc. A full programme of activities is also provided, for evening entertainment. Volunteers work for three of the four weeks in September or October and may then spend their final week travelling around the island. An orientation weekend must be attended by all volunteers and applications should be in by March. The all inclusive cost is approximately £650 including £100 for pocket money.

Further information from: Britain–Cuba Resource Centre, José Marti International Work Brigade, Latin America House, Priory House, Kingsgate Place, London, NW6. Tel. 071 388 1429.

KENYA VOLUNTARY DEVELOPMENT ASSOCIATION.
Workcamps are organised in Kenya's rural areas where the organisation encourages volunteers from Africa and abroad to

work with the local community on rural development projects. Emergency projects are also organised in response to local disasters. Recent work has included irrigation, re-planting and goat and hen rearing. Volunteers must be eighteen or over and work six hours a day. Community entertainments are organised during free time. No payment is given but food and accommodation are provided, and volunteers must make their own way to the site. Workcamps last two to three weeks in April, July, August and December. There is a registration fee of $160 for one camp.

Further information from: The Director, KVDA, P.O. Box 48902, Nairobi, Kenya.

THE MONKEY SANCTUARY.

Established in 1964, this centre has received worldwide acclaim for its successful breeding and conservation work. Volunteers aged twenty-one and over, are involved in all aspects of the sanctuary, from food preparation and cleaning to spending time with the monkeys in their enclosures. Food is usually vegetarian and volunteers are encouraged to take their musical instruments to provide entertainment. Stays are for two weeks or more, all year round and full-board and shared accommodation is provided. Approx. £5 per week pocket money paid to volunteers staying several weeks.

Further information from: The Head Keeper, The Monkey Sanctuary, Murrayton, Looe, Cornwall, PL13 1NZ. Tel. 05036 2532.

WORKING WEEKENDS ON ORGANIC FARMS.

Founded in 1971 and run by organic food enthusiasts, this organisation aims to provide willing hands for organic farmers who need labour to replace the multitude of chemical, herbicides and machinery used in conventional food production. Volunteers, affectionately known as WWOOFERs, work weekends as a non-financial exchange: you work in the fields and the farmer provides food, accommodation and transport from the local railway station if required. An annual subscription fee of £6 gives you a list of farmers needing helping hands, updated every two months. You need to be sixteen or over. If you enjoy the farmwork at weekends in the UK and Ireland, then WWOOF is in contact with similar organisations throughout Europe, and beyond, at which volunteers may spend a week or more.

Further information from: WWOOF, 19 Bradford Road, Lewes, Sussex, BN7 1RB. Tel. 0273 476286.

National Information Addresses

HOLIDAY CARE SERVICES.

A national charity established in 1981 as a centralised specialist information service for people whose needs make holidays difficult or impossible. In particular it helps those elderly and frail, those with a disability or learning difficulty, or those who have special financial problems through single parenthood, widowhood or long-term unemployment. Holiday helpers are used and always needed. Tourism for All is their successful campaign for the 90s and the information they can provide for those with special needs is well worth requesting.

Further information from: 2 Old Bank Chambers, Station Road, Horley, Surrey RH6 9HW. Tel. 0293 774535 Fax. 0293 784647

NATIONAL COUNCIL FOR VOLUNTARY ORGANISATIONS.

Founded in 1919, the Council aims to act as a central agency for the promotion of voluntary services throughout Britain. While the Council cannot find suitable voluntary work for individuals, for 85 pence it will send a comprehensive list of addresses of the local and regional offices with which it is in contact. Volunteers may then contact their regional/local organisation who will be happy to provide information on local volunteer requirements.

Further information from: National Council for Voluntary Organisations, 26 Bedford Square, London, WC1B 3HU. Tel. 071 636 4066.

THE VOLUNTEER CENTRE UK.

An independent charity funded by charitable trusts, this is a national agency looking at all aspects of volunteer involvement in Britain. For prospective volunteers the centre has an information bank and produces regular publications.

Further information from: The Information Unit, The Volunteer Centre UK, 29 Lower King's Road, Berkhamstead, Hertfordshire, HP4 2AB. Tel. 044 27 73311.

NB: When applying to projects overseas, remember to enclose an international reply coupon. For projects in this country enclose a stamped addressed envelope.

Chapter 9

Wildlife/Ecological Holidays

For many people the animal kingdom and its habitats seem so far removed from our everyday existence that the desire to experience 'real nature' has turned wildlife holidays into a growth market. The recent upsurge in environmental awareness has also emphasised the importance of conserving flora and fauna. Not only has interest increased but the potential revenue of wildlife and associated tourism has been recognised by governments and landowners who previously attached little value to wilderness areas. One paradox of this interest is the increasing pressure on fragile habitats.

Wildlife and ecological holidays represent, to a large extent, a good example of 'green' tourism. In the US, and increasingly in the UK, the term 'Ecotourism' has been used to describe, 'travel to relatively undisturbed natural areas to view and study the flora and fauna, often combined with understanding the local indigenous culture.' It certainly involves a desire to know more about the natural world and thus fits in with the increasing trend of greater environmental awareness.

Such holidays are by no means new. Some of the early tourist holidays involved trips to view wildlife or spectacular natural phenomena. Many centred on hunting and shooting and the benefit to the wildlife was often, at best, questionable. Nevertheless, such holidays do represent one end of the 'Ecotourism' spectrum. The fisherman hiking to a remote lake to fish and 'communicate' with nature is arguably following a reasonably environmentally friendly pursuit. More complex is the action of, for example, the Zimbabwe government, who are actively promoting 'Big Game' shooting safaris of elephant and buffalo, in conjunction with the World Wide Fund for Nature. They argue that their game management policies are sufficiently successful to necessitate culling, to prevent over-population and habitat destruction in national parks and game reserves. Using tourists to do the culling and thus raising revenue allows for more resources to be devoted to overall wildlife conservation.

Of course people have their own views on the moral rights and wrongs of hunting, fishing and shooting as recreational activities. It can be reasonably argued that such activities, when properly managed, can sustain and support wildlife. It's in the hunter's interest to ensure that the prey population is properly managed and does not disappear. And, provided such tourist activities are handled in a way compatible with sustainable principles, they may have a positive contribution to make to both tourism and the wildlife.

The main holidays for consideration in this chapter are those dealing with observation, viewing and study of wildlife. These are represented by a considerable number of operators, although as yet they do not constitute a particularly large percentage of the industry as a whole. One of the reasons for this is that safaris, which constitute a significant portion of the wildlife holiday market, usually cost more than people want to spend on their annual holiday, averaging out at around £1,700–£2,000 for ten to fourteen days in one of the African countries. That said though, the market is definitely expanding.

What is a Wildlife/Ecology Holiday?

The range of wildlife/ecology holidays available is quite considerable. Birdwatching tours continue to be popular and nowadays the wildlife 'options' range from studying Darwin's views on natural selection in the Galapagos Islands, to viewing whales off the coast of California, or studying the ecology of gorillas in the mountains of Rwanda in Central Africa.

One of the most well-known types of holiday is the traditional safari, usually offered in one of the East African countries, although the term 'safari' is now being applied to any sort of wildlife holiday which involves a variety of animals viewed in different locations. Among some, there is also a feeling that the more remote the safari location, the better the holiday should be. This brings with it a whole set of problems concerning the development of 'non-tourist' areas.

Safaris today are offered all over the world, from Canada, Iceland and Greenland, to East Africa. While such holidays provide tourists with a valuable insight into the fauna of the land, there is potential for damage to often fragile, sensitive and relatively undisturbed environments. Particular care is required to ensure such holidays are undertaken in a way which will not damage the environment. Companies offering safaris range from large, well established operators such as Cox and Kings and Naturetrek, to smaller organisations like Grass Roots and Papyrus Tours. The main difference is that smaller companies tend to be run by individuals or couples who, as a result of some time spent in the country of interest, have set up their own tours.

Birdwatching also draws a significant number of tourists. Increasingly companies are branching into areas such as South America, India and the Arctic countries, as well as more traditional destinations such as Europe and East Africa. Generally operators tend to be fairly small, more in the mould of 'special interest' companies rather than mainstream chains. These include concerns such as Cygnus and Ornitholidays, as well as conservation organisations such as the Wildfowl and Wetlands Trust.

The Arctic regions are also important to whale and dolphin watching, as are the Pacific and Atlantic coasts of North America and the seas around Hawaii, the West Indies, Sri Lanka and New Zealand. Commercial viewing of whales and dolphins is on the increase and fortunately some operators are now listening to the advice of conservationists concerned about tourism intrusion. As yet, the number of companies involved with whale and dolphin watching is relatively small. Those that do offer trips such as Arctic Experience and the Skye Environmental Centre, adhere to strict guidelines either set down by authorities such as the Marine Conservation Society or responsibly imposed by and on themselves.

Coral reefs are an invaluable source of wildlife and underwater viewing either by diving or in glass-bottomed boats. Such boat excursions often play an incidental role within the framework of a traditional beach holiday to places such as the Red Sea, Australian Great Barrier Reef, the Cayman Islands or the Seychelles. Diving holidays are now a reason to travel within themselves and are offered by large and specialist companies alike. Twickers World and Equinox Travel are two examples covering the Maldives, Thailand, the Red Sea and the Caribbean.

The Amazonian rainforests, although much publicised by the media, have yet to receive much attention from the tourist industry. Twickers World offer trips to the Amazon as do Cox and Kings, both educating their clients about the need to conserve this area. Similarly, Passage to South America offers specialist expertise and puts together programmes which show the Amazon and other wonders of the rainforest areas, based on sustainable policies.

This is an indication of the range of wildlife/ecological holidays and the type of companies which offer them. What prospective wildlife holidaymakers should remember is that, increasingly, the point of such tours is to *learn* about the animals and ecosystem, rather than just observing for the sake of adding to the list of 'things seen'.

How Green are Wildlife/Ecological Holidays?

The attractions which lure people to Africa or through Amazonian rainforests or around a sanctuary such as Bird Island in the Seychelles, all stem from the natural environment. The fauna and flora are indigenous to their surroundings; tourists are not. If tourists and tour companies do not respect and integrate with such surroundings, they will end up destroying the very resource which they have come to learn about and enjoy.

Sustainable tourism is, or at least should be, at the centre of such holidays: learning about, integrating with, respecting and understanding the environment or ecosystem you visit. It is no coincidence that a leading company in the field of African safaris is actually called *Eco*Safaris. They have signed an agreement with the Zambian government which allows them sole responsibility for the running, administration and conservation of one of Zambia's national parks, Kasanka.

Wildlife/Ecological Holidays

Safaris are a significant part of the travel market, representing at least forty operators from the UK. One serious side effect of safaris, however, is land erosion. National parks cannot sustain the wear and tear of safari vehicles without implementing compensatory measures. Fortunately, steps are being taken – such as roads being closed off periodically – to ensure repairs can be carried out. Nevertheless, some responsibility remains with the tourist who should not ask drivers to leave the road.

One of the most common problems on wildlife holidays arises from tourists who adopt the attitude that they must see as many different types of animal or bird as possible. Reserves and parks are not glorified zoos, and wardens, operators and drivers have no control over animal behaviour. Research at Amboseli National Park in Kenya discovered that almost 80 per cent of tourists kept within a fifteen square kilometre area when viewing game. This was attributed to the fact that tourists were looking for specific animals – lions and cheetahs – which were known to be in the area. Since neither species was present in great numbers the situation arose where single animals had fifteen to thirty vans surrounding them.

The enjoyment of wildlife holidays should come from seeing animals in their natural habitat, regardless of what species they are and the amount you see. Inevitably you might feel disappointed if you don't see a rhino, or if a rare bird doesn't make an obligatory appearance, but you can at least be satisfied that these animals do exist and that you're lucky enough to be experiencing their environment. Admittedly, watching wildlife is better than destroying it, but in keeping animals alive they must be able to live as 'natural' a life as possible.

Poaching is a further problem. Today poaching still goes on even in the most advanced societies. Ivory remains a valuable commodity to those intent on selling it and animal skins and furs still fetch a high price on some of the world's busiest high streets. Although action such as President Moi's public burning of ivory in Kenya in 1989 made it clear that poaching will not be tolerated, tourists can still help put an end to the trade by refusing to buy animal artefacts. They can also contribute to societies such as Elefriends or Save the White Rhino, which strive to protect particular species. This can be done by direct contribution or by booking a holiday through a company such as Twickers World who contribute to the World Wide Fund for Nature.

Intrusions of various sorts apply to all wildlife holidays. The consequences are often too subtle for tourists to realise but do have far-reaching effects. Damage can be done by erosion of vegetations by the passage of feet or vehicles; disturbing animals away from nesting grounds, thus leaving eggs open to predators; disturbing animals from their territory, thereby separating sexes and frustrating breeding and separating parents from offspring before the natural 'bonding process' has been completed. In this last instance – which usually happens when over-eager tourists in their vans or boats plunge into the middle of a group of animals – the result for some species is that parents and young are not reunited and offspring perish due to lack of food or fall victim to predators. Tourists and operators must adhere to any rules stipulating the distance allowed between boats/vehicles and animals and the number of vehicles which should be present at any one time. Controls have been successfully applied in Rwanda, for example: during trips to see the mountain gorillas the number of tourists per day is limited and the amount of time spent with the gorillas is kept to a minimum.

Animals should also not be fed. The effects of disregarding this advice have been seen in the USA and Canada, particularly in relation to the North American Black Bear. For years after parks such as Yellowstone and Yosemite were opened, tourists were allowed to tempt bears closer with hand-held food. As the bears became accustomed to this, they began to expect food from tourists. When they were denied it they reacted violently, sometimes breaking into vehicles and cabins. In cases where human injury was reported it was the bears who had to suffer and not the people. Nowadays most parks display signs warning you not to feed the animals, but responsibility ultimately lies with the tourist.

Wildlife/Ecological Holidays

PRACTICALITIES

If you are considering a wildlife holiday there are one or two points to take into consideration. First are your reasons for going. Are you really interested in seeing animals in their natural habitat, or learning about the ecology of an area? Or is it just a romantic *Out of Africa* notion of time spent on the African plains or sailing up the Amazon that attracts you to such a venture? Remember, many of these tours are highly specialised with usually ten to fourteen days dedicated to observing wildlife. To get the most out of such an experience you need to be really interested in what you are doing. Some operators will incorporate a couple of days by the pool or on the beach, but others like to totally immerse themselves in their surroundings, often in fairly remote places. The last thing you want is to spend half your time looking forward to returning to the city.

Also worth thinking about is with how many people you wish to travel. It is often the case that once 'on location', many tourists find themselves wishing they had come with a smaller tour. The difference between travelling with a bus-load and travelling with maybe just four or five others can alter your whole perspective on a holiday. What may seem like a 'sociable' idea at home does not always transfer smoothly to different parts of the world. You may have to pay slightly more to go with a smaller group, but it can be worth it.

The final consideration deals with the weather. Wildlife holidays often take in countries where the weather is extreme – incredibly hot or freezing. In Africa you'll probably spend most of your time just a few miles off the equator. Although cool at night, during the day temperatures become not just hot but intense. For some people this is unbearable. If so, is your interest strong enough to compensate for a certain amount of discomfort? Alternatively, trips to the Arctic Circle can find you shivering with cold. Do ask yourself before you book if you are prepared to spend your precious time on holiday putting up with physical discomfort.

WILDLIFE/ECOLOGY HOLIDAYS AVAILABLE

The companies listed here are those who have shown a genuine awareness of their role in protecting the environment and whose

Arctic/Antarctic Holidays

ARCTIC EXPERIENCE LTD.
Arctic Experience offer a variety of wildlife tours to Iceland, Greenland, Spitsbergen, Norway, Sweden, Scandinavia, Russia and Canada. Birdwatching, 'polar bear adventures', grizzly bear and whale watching trips are all available. Arctic Experience believe that, 'tourism can benefit conservation by promoting a better understanding of the many threats to the natural world', and that it can 'provide a strong economic argument for protecting wildlife and wild places.' The company contributes to the Wildlife and Wetlands Trust and the Whale and Dolphin Conservation Society. Between January and March of each year they also organise a series of slide shows around the country, visiting no less then twenty-five British cities. Prices vary, but the fourteen-night 'BirdWatcher's paradise' tour to Iceland costs roughly £1,700, while the nine-night 'Belugas & Bears' tour in Canada costs roughly £1,850. Arctic Experience also offer trips to St Kilda and other Scottish islands. Prices range from £800 to £2,800.

Further information from: 29 Nork Way, Banstead, Surrey, SM7 1PB. Tel. 0737 362321 Fax. 0737 362341.

ARCTURUS EXPEDITIONS.
Previously Erskine Expeditions, the company runs tours to Spitsbergen, North Greenland, Canadian Arctic, the Falkland Islands and the Arctic in general. Muskox, caribou, polar bears, walrus, snowy owls, Sabine's gull and the 'spider plant' are just a few of the highlights. Hike over the tundra and observe wildlife in the summer, go dog-sledging and ski-touring in the spring, or simply cruise around the fjords in Spitsbergen in a small ship, disembarking each day for walks to see the wildlife.

Further information from: Arcturus Expeditions, 80 Caxton End, Eltisley, Huntingdon, Cambs. PE19 4TJ.

FRED OLSEN HOLIDAYS.

Included in Fred Olsen's 'World of Difference' brochure are a selection of cruises entitled 'Voyages of Discovery'. Alaska and Antarctica are both featured using the new *MS Frontier Spirit*, which has been designed with all the latest technology in environmental terms. The cruises range from £1,770 to £9,420. £10 is donated to the RSPB for every passenger carried. Fred Olsen say that, 'the expertly and meticulously planned voyages along less-travelled, but immensely rewarding, routes are designed for those who combine an interest in the natural world with a love of exploration, excitement and first-hand experience.'

Further information from: Crown House, Crown Street, Ipswich, Suffolk, England, IP1 3HB. Tel. 0473 230530.

OCEAN CRUISE LINES.

As part of an extremely wide-ranging brochure, Ocean Cruise Lines are offering cruises to Antarctica and the Falklands on board their luxury liner the *Ocean Princess*. Ocean Cruise Lines are not the only company to offer trips to this part of the world but they are one of the only companies who are working in conjunction with Lars-Eric Lindblad, who pioneered trips to Antarctica in the early 1970s on board his ship the *Lindblad Explorer*. Although the *Lindblad Explorer* is not used by Ocean Cruise Lines the company is benefiting from Mr Lindblad's expertise, ensuring that their trips are of as least damaging a nature as possible. Prices range from £1,350 to £4,600.

Further information from: 10 Frederick Close, Stanhope Place, London, W2 2HD. Tel. 071 724 7555; Fax. 071 402 0490.

Birdwatching

BARN OWL TRAVEL.

Barn Owl Travel have been in operation since 1973 and are specialists in the birds of the British Isles. Other destinations include Austria, France, Iceland, Israel, Menorca, the Spanish

Pyrenees, and the Netherlands. Holidays are arranged for individuals and for groups of up to eight and many are led by the director whose main interests are birds and butterflies. Weekends in Kent, boat trips on the River Medway, full holidays in selected areas of the UK and Europe and some safari work in Africa are all available. A Green Flag International member.

Further information from: 21 Heron Close, Lower Halstow, Sittingbourne, Kent, ME9 7EF. Tel. 0795 844464.

BIRD HOLIDAYS

This independent company was set up to raise funds for the Royal Society for the Protection of Birds. Expert leaders, mainly RSPB staff, lead tours to Corfu, Crete, Hong Kong, Iceland, Majorca, Spain, Tunisia and Turkey as well as to RSPB reserves in Britain. Accommodation is normally in 3- or 4-star hotels with daily excursions by coach or on foot. Prices range from £500 for one week in Tunisia to £1,200 for two weeks in Iceland. Staff and leaders are involved in action to protect sites and influence conservation attitudes in the countries they visit.

Further information from: Dudwick House, Buxton, Norwich, NR10 5HX. Tel. 0603 278296.

BIRD WATCHING CRUISES.

This company offer birdwatching cruises around the Scottish Islands on board the MV Monaco. Run by the manager of the NCC Rhum Sea Eagle project, featured on the BBC's *Wildlife on One*, tours offer clients the opportunity to view the abundance of birdlife in this beautiful part of Scotland. Trips round Mull and on the Caledonian Canal are very popular.

Further information from: Amelia Dalton, Manor Farm, Chaigley, Clitheroe, Lancs, BB7 3LS. Tel. 025 486 591/461.

BIRDING

John and Robbie Gooders run birdwatching trips to a variety of destinations including Cyprus, Corsica, Yugoslavia, Portugal,

Spain, Poland, Austria, Hungary, Jordan, India, Nepal, The Seychelles, Canada and the Gambia. Groups are restricted to sixteen plus two well-travelled and experienced leaders.

Further information from: Birding, Penteau House, Winchelsea, East Sussex, TN36 4EA. Tel. 0797 223223.

BIRDQUEST LTD.

The range of destinations offered by Birdquest is as impressive as their brochure. Thailand, Madagascar, Turkey, Siberia, Japan, Zimbabwe and Australia are all included in the brochure (among 35 countries). These are tours for real birdwatching enthusiasts and the directors state that all you really need is 'a desire to watch birds for days on end, a certain sense of adventure and an ability to get on reasonably well with others in the group.' Sample prices are nineteen days in Canada for £2,370; eight days in Hungary for £690. Professional leaders and low-impact facilities.

Further information from: Two Jays, Kemple End, Stonyhurst, Lancashire, BB6 9QY. Tel. 0254 86317 Fax. 0254 86780.

CYGNUS WILDLIFE HOLIDAYS.

Cygnus Wildlife offer bird watching holidays all over the world, from Egypt to India, Austria to Brazil, Hungary to Hong Kong. Their tours manager honestly states that, 'we do not claim to be doing any more than arranging birdwatching holidays', as opposed to making a conscious effort to contribute to local economies or encourage local contact. He does make the valid point, however, that simply due to the nature of their trips, 'conservation is inextricably part of every tour we operate' and that since their holidays 'frequently involve visiting reserves and paying entrance charges, this will inevitably assist the local conservationist.' Mention should also be given to Cygnus' 'protracted lobbying of the Cyprus government to try and enforce a shooting ban on spring migrants.' Although Cygnus do not have any formal written policy on conservation, their dedication to their subject ensures the responsible execution of their trips. Prices range from £500 to £2,700 for eight to twenty-two-day trips.

Further information from: 57 Fore Street, Kingsbridge, Devon, TQ7 1PY. Tel. 0548 856178.

ISLAND HOLIDAYS

Scottish-based Island Holidays started five years ago offering only the Falkland Islands. Today they have around ten destinations with Christmas on Christmas Island (£1,720 for eleven days) being their latest offer. Closer to home they will be operating to Orkney and Shetland this summer with small groups led by partner Bobby Tulloch, Shetland's best known naturalist. A member of Green Flag International, Island Holidays operates at the top end of the market. They also work in conjunction with *Wine Dark Sea*, offering natural history/archaeological holidays to Crete.

Further information from: Ardross, Comrie, Perthshire, PH6 7JU. Tel. 0764 70107 Fax. 0764 70958.

ORNITHOLIDAYS.

Established in 1965, Ornitholidays was the first corporate member of the Royal Society for the Protection of Birds. Their worldwide programme is extensive covering Corfu and Cyprus to Venezuela and Yugoslavia, Assam and Sikkim. All tours are escorted by experienced British ornithologists and often include excursions by bus or on foot to lagoons, salt pans, meadows and mountains, always exploring those areas renowned for their birdlife. Prices vary, but a seven-night trip to Spain costs about £655.

Further information from: 1–3 Victoria Drive, Bognor Regis, West Sussex, PO21 2PW. Tel. 0243 821230.

PEREGRINE HOLIDAYS.

Peregrine Holidays have been involved with birdwatching/natural history holidays and conservation for over twenty-five years. Spain, Egypt, Holland, Costa Rica, Norway, Czechoslovakia, Poland, Morocco, France, Hungary, Austria, New Zealand, the Gambia, several other African countries and some in S.E. Asia are all on offer and the company is also a specialist in Greece. Peregrine's policies are reflected in their recent action taken against the Cypriot government (along with Cygnus Wildlife) when, for the second year running, they cancelled all holidays to Cyprus in protest at the Cypriot Government's failure to adhere to the Berne Convention on migratory birds. The company are concerned only with the observation and conservation of birds and, as such, follow set guidelines when running their tours. Prices range from £550 to £1,250 for two weeks.

Further information from: 41 South Parade, Summertown, Oxford, OX2 7JP. Tel. 0865 511642.

SUNBIRD.

Voted the best of the major birdwatching holiday companies in a survey by *British Birds* magazine, Sunbird offer thirty-three tours which cover all continents. All are accompanied by professional leaders and group sizes are restricted in order to minimise intrusion and allow for a more personal service. Sample prices are £880 for seven days in Poland, £1,695 for sixteen days in Kenya.

Further information from: PO Box 76e, Sandy, Bedfordshire, SG19 1DF. Tel. 0767 682969 Fax. 0767 692481.

WILDFOWL AND WETLANDS TRUST.

Established in 1946, the Wildfowl Trust has eight centres throughout the UK, focusing on wetlands and wetland wildlife.

Their primary aim is to educate their clients about these areas by means of four activities: 'Conservation, Education, Research, Recreation'. They state that, 'wetlands are perhaps the easiest habitat in which to mix wildlife and people as the water acts as a "barrier of trust" between them . . . The mixing of people and wildfowl, however, is achieved by manipulation . . . Visitor interest is largely channelled into "honeypots" in order to leave other areas alone. While some may find it hard to accept, all wildlife is managed in one way or another.' Despite this sobering thought, or maybe because of it, the Trust has been so successful as a conservation organisation that 'numbers of wildfowl have increased so much that consideration is now having to be given as to whether control of certain species is needed.' They are one of the foremost operators in the green tourism field and their work is of international importance.

Further information from: Slimbridge, Gloucester, GL2 7BT. Tel. 0453 890503.

Other birdwatching companies include BUFO VENTURES LTD: 3 Elim Grove, Windermere, LA23 2JN. Tel. 09662 5445, for Nepal and India. FLAMINGO HOLIDAYS: 59 Hunger Hills Drive, Leeds, LS18 5JU. Tel. 0723 891715, for Siberia, Nepal, Gambia, Spain and Majorca. GOSTOURS: 109 Hammerton Road, Sheffield, S6 2NE. Tel. 0742 334171, for Majorca, the Greek Islands, Egypt, Israel, Cyprus and Istanbul. NATURETREK, Chautara, Bighton, Nr. Alresford, Hampshire, SO24 9RB. Tel. 0962 733051.

British Wildlife

AIGAS FIELD CENTRE.
Aigas Field Centre offers holidays and island expeditions concentrating on the wildlife of Scotland's Highlands and Islands. Peregrines, eagles and divers are common sights during the summer, as are red deer which can be observed in the glens. Shunning the idea of 'collecting animals and birds' in mental lists, Aigas offer their clients an opportunity to study species over a week, learning about their habits and behaviour. Courses based on natural history, archaeology/history, garden tours, etc. Full board or B&B. Prices start at £375 for a week.

Further information from: AIGAS, Beauly, Inverness-shire, Scotland, IV4 7AD. Tel. 0463 782443.

GLENLYON WILDLIFE SAFARIS.
Donald Riddell organises trips in one of Scotland's glens primarily for photographers of wildlife, but anyone is welcome. Eagle, peregrine, ptarmigan, deer, badger and grouse are just a few of the highlights. Landrover or walking options. Day safaris from £15–£65.

Further information from: Donald Riddell at Dornoch Lodge, Glen Lyon, Aberfeldy, Scotland, PH15 2NH. Tel. 0877 877235.

SKYE ENVIRONMENTAL CENTRE.

Wildlife is one of the main attractions at the Skye Environmental Centre. The centre was formed in 1987 after amalgamating with the Isle of Skye Field Centre which had been running holidays since 1984. The range of wildlife which can be observed on this and other visited Hebridean islands is impressive and highlights include the peregrine falcon, whales, porpoises, the golden eagle and otters. Also available for use is the Environmental Resource Centre which, as well as containing maps, periodicals and books on natural history, houses a comprehensive geological data-base linked to the National Scheme for Geological Site Documentation and acts as a recording unit for the Highland Biological Recording Group. New for '92 are holidays on the Isle of Rhum. Prices range from £95 to £445 for a long weekend to a week.

Further information from: Isle of Skye Field Centre, Broadford, Isle of Skye, IV49 9AQ. Tel. 04712 487.

SEALIFE CRUISES.

In conjunction with the 'Whale & Dolphin Conservation Society', Sealife Cruises undertake observation and survey work in the seas around the island of Mull. Full-board and accommodation are provided. Daily wildlife trips are also available. Prices range from £360 to £620 for a five- or seven-day package, though prospective clients must first get themselves to Mull at their own expense. Their aim is to study the migratory patterns and existence of indigenous pools of killer, minke and pilot whales. They are run by Richard Fairbairns who has lived on Mull for fourteen years.

Further information from: Sealife Cruises, Quinish, Dervaig, Isle of Mull, PA75 6QT. Tel. 06884 223 Fax. 06884 383.

WESTERN ISLES SAILING AND EXPLORATION CO.

Based on the sailing boat the *Marguerite Explorer*, the company offers various natural history and birdwatching trips to Ireland, Skye, the Hebrides and St Kilda. Whale and dolphin watching, themed cruises, and a high standard of accommodation. Individual berths from £300 a week.

Further information from: Prospect House, Hollands Road, Haverhill, Suffolk, CB9 8PJ. Tel. 0440 702454 Fax. 0440 62593.

Safaris

ABERCROMBIE AND KENT TRAVEL.

Few tour companies can compete with Abercrombie and Kent when it comes to constructive efforts to conserve the environments which they visit, notably those in East Africa. The company have been organising safaris to Kenya and Tanzania for over thirty years and have used their success to initiate various environmental movements throughout Africa. The most recent of these, a charity called 'Friends of Conservation', is centred on Kenya's famous Masai Mara and aims to 'work closely with the Narok County Council to establish a well-run and safe reserve; to totally eliminate poaching and to educate the tourists and locals alike of the advantages of treating the reserve and its wildlife with the respect they deserve. Also to educate the locals of the benefits of conserving wildlife as a valuable asset which, properly cared for through tourism, can create a large income for the area's residents.' This is just a brief insight into the work of Abercrombie and Kent and their associated charity. It goes without saying that their tours are highly recommended. Prices range from £1,500 to £9,750 for ten- to twenty-eight-day trips (for example, British Columbia whale watching tour £1,565 for eleven days).

Further information from: A&K, Sloane Square House, Holbein Place, London, SW1W 8NS. Tel. 071 730 9600.

AFRICA TRAVEL CENTRE.

Offers low cost, no frills safaris throughout Africa, including walking safaris and special interest theme safaris. Also 'Africa by train' adventures. A realistic approach to their holidays is given and integration with locals is encouraged. Sample price, twenty-five days in Rwanda, Zaire and Tanzania, £510.

Further information from: 4 Medway Court, Leigh St, London WC1H 9QW. Tel. 071 387 1211 Fax. 071 383 7512.

Wildlife/Ecological Holidays

CARA SPENCER SAFARIS.

Cara Spencer has been escorting safaris around Kenya and Tanzania for over twelve years. Her brochure describes her trips as, 'tailor-made holidays to suit all tastes with opportunities to get off the beaten track and experience the best of Africa.' As well as belonging to Zoo Check, Elefriends, Greenpeace, the East African Wildlife Society and the National Trust, she also works in conjunction with Virginia McKenna of *Born Free* fame, who now dedicates her time to conservation. Cara, as a born and bred African, places a great emphasis on what the locals feel. Pre-departure briefing notes are given and her *Born Free* Safaris stress 'green themes' throughout. £100 donated to Zoo Check for every safari booked. Personal service is combined with a full education in all things African on Cara Spencer tours, during which the director will do her utmost to 'avoid the infamous convoys of mini-buses, the touristy lodges and cowboy safari drivers', opting instead to encourage clients to 'appreciate the local African culture', and to help them 'understand that what may appear uncivilised in our eyes is right and appropriate to these people who have always lived in harmony with animals and nature'. Trips are also organised to Zaire, Zimbabwe, Malawi and Botswana. Prices range from £1,775 to £2,200 for two weeks.

Further information from: 11 Cavendish House, Eastgate Gardens, Guildford, Surrey, GU1 4AY. Tel. 0483 574437 Fax. 0483 66761.

ECOSAFARIS.

Following in the footsteps of Norman Carr, one of the original big game hunters and game wardens who pioneered Luangwa safaris, EcoSafaris are leading the way today in creating the new 'ecotour'. Take a 'walk on the wild side' through Africa, Asia, South America, the Arctic and Europe, learning about the wildlife, birdwatching, trekking in the Himalayas, going on the 'panda trail' or exploring the Nile. EcoSafaris' range of trips is extremely impressive as is their acute awareness of the dangers of tourism to the environment. Not only do they work in association with and support the Wildlife Conservation Society of Zambia, the Royal Society for the Protection of Birds, the World Wide Fund for Nature, and the Royal Society for Nature Conservation, they have gone a step further and actually bought a park, Kasanka, which is now run under their administration in an effort to educate clients further about the African wilderness. EcoSafaris believe they have, 'the highest degree of specialist knowledge of Zambia available today'. Prices range from £1,000 to £3,050 for sixteen to twenty-five days.

Further information from: 146 Gloucester Road, London, SW7 4SZ. Tel. 071 370 5032/3.

GRASS ROOTS TRAVEL LTD.

Grass Roots offer clients both escorted and independent travel in Kenya, Tanzania, Namibia, Botswana and Zimbabwe and to the mountain gorillas in Rwanda and Zaire. Escorted groups are kept to ten people. As members of the East African Wildlife Society the company uses ground operators whom they feel 'are genuinely aware of the dangers of over-exploiting the environment' and companies 'whose drivers/guides have an in-depth knowledge of the areas they are working in, and are sensitive to the needs of wildlife.' Grass Roots educate clients about the 'inter-relationships between climate, soils and vegetation and the impact these can have on wildlife distribution; the relationship between the natural environment and people, especially where rapid population growth is causing severe land shortages; the significance of tourism to economies via foreign exchange earnings and providing employment; and by talking with drivers/guides in order to gain insight into the stresses and strains of everyday life in local communities.' They work in the belief that, 'with planning, the conflict (between the promotion of tourism and the preservation of culture) can be minimised and a valuable understanding can be achieved.' Grass Roots are also one of a small handful of companies who offer tours to Indonesia where they follow the same principles, endeavouring to 'provide a greater insight into the cultural background of the areas visited by explaining *en route* the ethnic characteristics; e.g. tribal groups, building styles, agricultural techniques, work patterns, clothing, religion etc.' Prices range from £1,200 to £3,000 for two weeks; £1,700 for three weeks in Indonesia.

Further information from: 8 Lindsay Road, Hampton Hill, Middlesex, TW12 1DR. Tel. 081 941 5753 Fax. 081 941 7178.

INTO AFRICA.
The game parks of Kenya, the Ngorongoro Crater and the Chobe Desert in landrovers. Gorilla Safari in Zaire and a Gorilla Trek are highlights. Local drivers and low-impact accommodation used. Average price £500 for two weeks not including flights. Overlanding for up to twenty weeks available.

Further information: 70 Windmill Road, Croydon, CR0 2XP. Tel. 081 683 4744 Fax. 081 689 0253.

OKAVANGO TOURS & SAFARIS LTD.
As their name suggests, this company specialises in tours to the Okavango: Botswana, Zimbabwe and Namibia. Tailor-made safaris all in true conservationist spirit are what Okavango specialise in. They believe that, 'in order to operate in . . . the areas in which we specialise, we have to be involved with the environment and its preservation, since we believe that it will only be the Garden of Eden which it is, if we look after it.' Okavango are members of the Kalahari Conservation Society, Okavango Wildlife Society and the World Wide Fund for Nature and state that they, 'promote traditional ways of life by preserving much that would otherwise die out; e.g. tracking game, making and using Mekoro (dug-out tree canoes).' 75 to 80 per cent of their turnover goes into the economy of the country in which the tour actually takes place. An interesting overland for '92 is one into Malawi and Zambia, and they also have extensions from safaris to Mauritius. Prices range from £1,310 to £1,900 per person.

Further information from: 28 Bisham Gardens, London, N6 6DD. Tel. 081 341 9442 Fax. 081 348 9983.

Other worthwhile safari companies to consider include ADVENTURE AFRICA: The Square, Ramsbury, Wiltshire, England. Tel. 0672 20569, for birdwatching and wildlife safaris in Kenya. AFRICA BOUND: 93 Chiswick High Road, London, W4 2EF. Tel. 081 994 9560. AFRICAN EXPEDITIONS: 2nd Floor,

Highgate House, 214 High Street, Guildford, Surrey, GU1 3JB. Tel. 0483 574939, for Zimbabwe and other parts of Africa. ART OF TRAVEL: 268 Lavender Hill, London, SW11 1LJ. Tel. 071 738 2038, for lake wilderness safaris, Zambezi river canoeing and walking safaris in Zimbabwe. Also for safaris/holidays in India, the Himalayas, and the islands of the Indian Ocean. GUERBA EXPEDITIONS: 101 Eden Vale Road, Westbury, BA13 3QX. Tel. 0373 826689 Fax. 0373 858351. Camping tours in several African countries highly rated. HOSKINGS WILDLIFE PHOTOGRAPHIC HOLIDAYS: Pages Green House, Wetheringsett, Stowmarket, Suffolk IP14 5QA. Tel. 0728 861113. OKAVANGO EXPLORATIONS: Regency House, 1–4 Warwick Street, London, W1R 5WB. Tel. 071 287 9672 Fax. 071 287 9673, for Zimbabwe, Namibia, Malawi, Zambia and Botswana's Okavango Delta. TONY PASCOE SAFARIS: The Old Rectory, Saltford, Nr. Bristol, BS18 3EU. Tel. 0225 873756, for safaris to Kenya, Tanzania, Zimbabwe and Botswana, led by an ex-game lodge manager who worked for nine years in Kenya's national parks. WILDLIFE EXPLORER LTD: 'Manyara', Riverside, Nanpean, St Austell, Cornwall, PL26 7YJ. Tel. 0726 824132, for Tanzania. WILDLIFE SAFARI: The Old Bakery, South Road, Reigate, Surrey, RH2 7LB. Tel. 0737 223903, for Kenya and Tanzania.

Wildlife/Ecology/Natural History

ACCESSIBLE ISOLATION HOLIDAYS.
Polar Bears, whales, Canadian-style safaris – there are several wildlife tours in Canada – a country that this specialist company knows well. Prices from £1,000.

Further information from: Midhurst Walk, West Street, Midhurst, W. Sussex, GU29 9NF. Tel. 0730 812535.

CHOOMTI TRAVELLERS.
Direct sell, host-based small company specialising in natural history tours in India. 'A Journey of India's Flora & Fauna' is a nineteen-day trip round some of the best bird and game

watching areas and national parks of the north – £1,350. Other trips available.

Further information from: 10 Todamal Lane, Bengali Market, New Delhi 110001 India. Fax. (91) 11 3322 810.

COX AND KINGS TRAVEL LTD.

Established in 1758, Cox and Kings have over 200 years of experience in travel. Their special interest tours, run as a non-profit making organisation, came about as a result of the recognition that 'responsible, environmentally-concerned tourism was perhaps the most powerful force available to preserve and to continue to observe some of the world's most threatened, yet beautiful, areas.' Trips take in the gorillas in Rwanda, parts of the Brazilian Amazon, Zaire and much more. Cox and Kings work in association with the Whale and Dolphin Conservation Society and are the official tour operator for the Programme for Belize. Their tours are strongly recommended. Prices vary considerably, but as examples, fifteen days in Crete costs about £1,100 while ten days whale-watching in Baja costs about £2,200. Three brochures are produced: India, Latin America and Environmental Journeys.

Further information from: St James Court, 45 Buckingham Gate, London, England, SW1E 6AF. Tel. 071 931 9106 Fax. 071 630 6038.

EUROPE CONSERVATION.

Europe Conservation is a non-profit making organisation which organises holidays and research programmes in parks and nature reserves throughout Europe. Their aim is 'the diffusion of knowledge about European natural resources to the Europeans themselves.' Among their past programmes has been a Field School of Avian Ecology and a Summer School of Alpine Ecology in the Italian Alps. Other activities include summer and winter 'schools' for organised groups at various centres throughout Italy, one of which is offered in association with the World Wide Fund for Nature in Italy. Europe Conservation's policy centres on, 'well programmed, low impact tourism in parks' which 'distributed throughout the year will be based on the utilisation of those structures already in existence and the development of activities recognised as having a low impact on the environment, also bearing in mind local community necessities.'

Further information from: Europe Conservation, via Fusetti, 14, 20143 Milan, Italy.

NATURETREK.

Naturetrek is now in its sixth year of operation, offering birdwatching and botanical treks and tours worldwide. Although some strenuous 'trekking' is involved in a selection of their trips (up Kilimanjaro for example), the emphasis tends to be on 'searching on foot for birds, mammals and plants.' The directors of the company who, along with other experts, escort the tours stress that, 'We do not aim to race from place to place in minibuses with the sole purpose of accumulating long lists of birds and mammals.' The result is that tours tend to be longer and move at a slower pace than others on the market and that there is more time for, 'photography, detailed wildlife exploration, relaxation and to take an interest in a region's people, culture, history and architecture.' They add that, 'We do not want our fascination in natural history to obscure other points of interest in the areas we visit.' Prices range from £790 for the seventeen-day 'High Atlas Mountains of Morocco' tour to £3,490 for a twenty-eight-day wildlife expedition, the 'Bhutan's Snowman Trek'.

Further information from: Chautara, Bighton, Nr Alresford, Hampshire, SO24 9RB. Tel. 0962 733051 Fax. 0962 733368.

PAPYRUS TOURS.

Established in 1984, Papyrus Tours offers, 'travel opportunities for wildlife enthusiasts' to Kenya and Tanzania as well as to the US. Florida was added in 1990 and France in 1991. East African tours take the form of safaris, while those to the US visit the Everglades, including the Everglades National Park, and coastal areas of Florida. In France, Papyrus visit a few of the 'many superb wildlife areas in this country of varied landscapes' which have survived 'in spite of the excesses of agricultural developments'. These include Rouen, St Amand, Tarn Gorge, Camargue and Les Dombes. Tours are, 'carefully planned to encompass a wide variety of habitats and, by avoiding short stays and limiting mileage, we aim to provide ample opportunity for tour members to absorb their experiences.' Prices range from £1,200 to £2,500 for fourteen to twenty-one nights.

Further information from: 9 Rose Hill Court, Bessacarr, Doncaster, DN4 5LY. Tel. 0302 530778.

PASSAGE TO SOUTH AMERICA.

This is a tour operator specialising in the provision of tailor-made holidays to South America for individuals. Groups can be arranged but they discourage them from visiting environmentally sensitive areas or indigenous communities. Local guides and locally-owned hotels are used. They have no set itineraries or 'packages'. Holidays are designed to take into account each traveller's priorities; and their consultants, who have first-hand knowledge of the continent, prepare itineraries in close communication with the passenger.

Their in-depth knowledge of the continent enables them to offer a very wide range of activities such as polo lessons, bird watching, river rafting, Galapagos cruises, jungle treks, Andean skiing, Patagonia camping and Amazon canoe trips. They also offer visits to all of South America's archaeological sites and can arrange expeditions to the isolated Indian communities.

Further information from: 41 North End Road, West Kensington, London, W14 8SZ. Tel. 071 602 9889 Fax. 071 371 1463.

REEF AND RAINFOREST TOURS LTD.

Reef and Rainforest Tours specialise in 'conservationist wildlife travel, particularly to Belize, but also to Argentina, Venezuela, Guyana, Trinidad and Tobago, Costa Rica, Christmas Island and Newfoundland.' The director is well acquainted with these areas and is often on hand to answer questions and explain more about the country. He states that, 'A percentage of the profits from our Argentine whale-watching tours goes to research projects concerning the southern right whale and we actively support and help the Whale and Dolphin Conservation Society. We are members of the Belize Audubon Society and support the Programme for Belize, a charity set up to preserve a huge tract of pristine rainforest in the Rio Bravo area . . . We also pride ourselves in giving a high degree of personal service.' Indeed, 90 per cent of their business is to Belize, offering individual itineraries for two people travelling together. Tours are on offer throughout the year with a total of six different itineraries. The company are now developing worthwhile programmes in Guyana and in Trinidad & Tobago, as well as offering excellent possibilities elsewhere – particularly in Argentina and Belize. Prices range from £1,400 to £3,300 for eleven to nineteen days.

Further information from: 205 North End Road, London, W14 9NT. Tel. 071 381 2204 Fax. 071 386 8924.

SNAIL'S PACE.
Snail's Pace is run as an agent for Mantra, Planned Management Travel. Greece and Yugoslavia are both visited and tours are suitable for beginners and old hands alike. Snail's Pace take an active interest in conservation, staying in small, local hotels where possible, visiting out of high-season, using village women's cooperatives in remote areas and submitting compiled reports to the appropriate authorities following their visits (Hellenistic Ornithological Society in Greece; Triglav National Park in Yugoslavia; and County Recorder in Britain).

Further information from: 25 Thorpe Lane, Almondbury, Huddersfield, HD5 8TA. Tel. 0484 426259.

SWAN HELLENIC.
Among the many tours offered by Swan Hellenic, including archaeology, art and architecture, art history, ballooning, cruises, wine tours and many more, there are special interest garden, natural history and safari holidays. The company was established in 1930 when it operated as a travel bureau and has since become part of the P&O group. Wildlife/ecology tours cover a wide range of destinations, including UK, Switzerland, Spain, Greenland, Kenya, China, the Galapagos Islands, Chile and Patagonia and all are accompanied by leaders and/or guest lecturers, all specialists in their field. Prices vary considerably but a botanical walking tour in Italy starts at roughly £1,275; a natural history tour of Malaysia and Borneo starts at roughly £2,400 and safaris to Africa start at roughly £2,400.

Further information from: 77 New Oxford Street, London, WC1A 1PP. Tel. 071 831 1616 Fax. 071 831 1280.

TRAVELLER'S TREE.

Traveller's Tree run unique tours to Bahia, Brazil and the Caribbean islands of Dominica, Grenada and Carriacou. Their main interest lies in natural history, but in pursuing that interest they offer more exploration of these places than any other operator. Indeed, they are the *only* company offering trips to Bahia, where clients are guaranteed a complete discovery of this fascinating Brazilian state.

Further information from: 116 Crawford Street, London W1H 1AG. Tel. 071 935 2291 Fax. 071 486 2587.

TROPICAL TRAILS.

Established four years ago, Tropical Trails offers lodge and camping safaris in Tanzania, with possible beach extensions in Mombasa and Zanzibar. The director of the company now lives in Tanzania on the slopes of Mount Meru and, as well as being able to offer some camping there, can invite clients to try out something slightly different at a new site in the Monduli Hills, staying in traditional Masai huts erected for guests' use. Tropical Trails operate climbing and walking safaris, as well as pioneering the way in 'Mountain Bike Safaris' in areas of the Rift Valley. Ngorongoro Crater, Lake Manyara, Kilimanjaro and Mount Meru are all areas known to the company's qualified guides, in addition to less frequented places such as Lake Natron, Oldoynio Lengi and Tarangire National Park.

Further information is available either directly from PO Box 6130, Arusha, Tanzania (Tel. Arusha 2365 Fax. Arusha 2365), or from the UK agent at The Beeches, Brunton, By Cupar, Fife, Scotland, KY15 4NB. Tel. 03377 257.

TWICKERS WORLD.

Twickers World have been in operation for twenty years, offering an increasingly large number of wildlife and natural history tours which now cover places such as the Galapagos Islands, the Amazon, Belize, South America, Costa Rica, the Arctic, Arabia, the Annapurna Sanctuary and the Falklands. Their enthusiasm for the conservation of the world's fauna and flora is impressive and some tours are run in conjunction with the WWF. Prices vary considerably but a four-night trip to Cairo costs about £450 while an eighteen-day tour to the Falklands costs about £2,900. For 1992 they are offering two of the first fleet re-enactment vessels which can take you on sections of a round-the-world trip, sailing to the Falklands, Tristan da Cunha, etc., from £1,065.

Further information from: 22 Church Street, Twickenham, TW1 3NW. Tel. 081 892 7606 Fax. 081 892 8061.

VOYAGES JULES VERNE.

Voyages Jules Verne offer an impressive number of unusual holidays throughout the world. As long-haul specialists they have put together approximately 100 different itineraries, with a good range of natural history/wildlife holidays. Greece, Spain, Canada, Yugoslavia, California (for whale-watching) and Papua New Guinea are all available for wildlife holidays, natural history, birdwatching and wildflower holidays, usually accompanied by an experienced tour leader. A bi-monthly newsletter gives details of all forthcoming tours. Prices vary, but ten days spent in France costs about £700.

Further information from: 10 Glentworth Street, London, NW1 6QG. Tel. 071 723 6556 Fax. 071 723 8629.

WILDLIFE TRAVEL.

Wildlife Travel was set up in 1987 to raise funds for RSNC, The Wildlife Trusts Partnership. It aims to organise travel programmes for naturalists from overseas to visit nature reserves and areas of natural beauty in the UK, and to take parties from this country to learn more about the culture and wildlife of other countries, especially in the Mediterranean. Staff and guides all have a deep concern for wildlife conservation, and the company is currently engaged in protecting sites in Crete and Cyprus by encouraging sympathetic tourism from people who care about the long-term future of the areas they are privileged to visit. Accommodation is normally in 3- and 4-star hotels with daily excursions by small coach or on foot. Prices vary from £450 for a week in Crete to £700 for ten days in Cyprus.

Further information from RSNC, Vigilant House, 120 Wilton Road, London, SW1V 1JZ. Tel. 071 931 0601 Fax. 071 828 6297.

WALSER VIAGGI.
This Italian company are now offering a six-night nature tour in the Piemonte region of northern Italy. Devised by the Vercelli Province Parks Department the tour, 'aims to show the discerning visitor something of the natural and cultural heritage of this diverse region.' The itinerary takes in Vercelli, Parco Lame del Sesia, Parco della Burcina, Valsesia, Lake Orta and Sacro Monte. Guest speakers and interpreters are always on hand, and every detail is taken care of. Prices are around £500, excluding air fare.

Further information from: Via A Manzoni, 9, 13100 Vercelli, Italy. Or contact Franca Rossi, 87 Fenwick Place, London, SW9 9NL. Tel. 071 274 7429.

Chapter 10

Alternative Holidays

Alternative holidays constitute a relatively small though fast increasing section of the travel market, born out of the growing desire for 'something different' and a holiday which offers more than the mass packaging of people to lie on a beach for two weeks. They are exactly what their description says – alternative, not mainstream. Into this category fall such options as: cycling holidays, farm holidays, naturist, study and personal development holidays, or indeed any holiday following a particular hobby or special interest.

As with any category which is given a set title there is a danger that the chosen label will be misinterpreted. In this instance the danger lies in the assumption that 'alternative' means something trendy, hippy, or for a radical few. This can lead to another misdirected assumption that 'trendy = green' and this is certainly not always the case.

The range of companies which come under the heading of 'alternative' is fairly extensive. In Britain companies such as Country Village Weekend Breaks and the Centre for Alternative Technology illustrate the use and value of tourism as an educational experience rather than a purely recreational one. Even in such instances, however, alternative holidays are only suitable for the 'good tourist', provided he/she takes the initiative to discover what is actually involved.

Since companies tend to be fairly small concerns, often run as family businesses by people who have become specialists in certain fields or areas of the world, their holidays are also fairly small-scale and often meet the criteria for sustainable tourism. Holidays can be loosely grouped into six broad categories although these are not mutually exclusive as there is some overlap between groups.

1. SPECIAL ACTIVITY, HOBBY AND PASTIME HOLIDAYS.

This type of holiday allows like-minded people to come together to enjoy the group pursuit of a common interest. Sporting, cycling, naturist, walking, pottery and woodworking holidays are all examples from this category. Often two or more pastimes can be combined on one holiday, allowing for a more varied trip (see Inntravel p.181).

2. LEARNING HOLIDAYS.

These are options which are proving extremely popular, particularly with the older generation, perhaps because this is a type of holiday which emphasises mental rather than physical exercise. Small groups travelling with small companies is the norm, learning a new subject in what is often the place of its origin. This also extends to companies such as Goodwill Holidays and the New Experimental College in Denmark, both of which specialise in integrating foreigners with the locals while at the same time attending some sort of study course.

3. HOLISTIC AND PERSONAL DEVELOPMENT HOLIDAYS.

This option continues the study theme. Companies such as Cortijo Romero offer a complete escape from city bustle to discover more about yourself than the average nine-to-five routine allows. Holidays cover a wide range of themes and often demand a high level of dedication.

4. ALTERNATIVE TECHNOLOGY, LIVING AND TRAVEL.

London Globetrotters organise interesting trips to places such as Nepal, Africa (the Sahara) and Papua New Guinea. Based on low-budget travel the bonus is not so much the cost as the experience of surviving in remote and distant lands on nothing but the basics.

Integration with the locals is essential to self-preservation on these trips, which bring even the most introverted out of their shells. A similar programme is offered by a German company, Alternativ Bus Reisen (Alternative Bus Travel). Journeys throughout the world are undertaken in a converted bus with self-sufficiency playing a central role.

5. HOMESTAYS.

Homestays specialise in certain areas of a country and are based on the client's participation and integration within a local community. The aim of these holidays is to live like and with the locals and to come away having gained an understanding of a way of life usually quite far removed from your own. Holidays either take the form of time spent with a tour in a certain country, such as Dick Phillips' running trips to Iceland, or time spent with a host family. Alternately a third branch of the market is 'home swaps', where families literally swap homes for the duration of their holiday. This also covers 'hospitality exchanges' which is, in effect, a combination of homestays, home swaps, and the traditional Bed & Breakfast. Travellers stay as 'one of the family' in a host's home and offer the same hospitality in return when the host is on holiday. Intercultural Education Programmes arrange for clients to spend extended periods of time living in a local household while on Goodwill Holidays' 'Meet the Russians' programme, clients stay in hotels but spend much of their time mixing with the local Russians.

6. UNUSUAL HOLIDAYS IN UNUSUAL PLACES.

This demands a category of its own. It covers one particular type of holiday or break which is becoming extremely popular, commonly called 'smokestack tourism' and involves touring some great cities to view the local industrial and environmental heritage. Major cities such as Bradford, Glasgow, Manchester and Liverpool are promoting this kind of tourism in an effort to conserve some of their industrial heritage and at the same time to promote their own image. Such visits can often also incorporate canal trips, city farms and alternative technology centres.

Other unusual holidays include trips which concentrate on the architecture of places visited, often in conjunction with some 'study' of fine art also, such as those offered by Inscape, Prospect or Martin Randall.

How Green are Alternative Holidays and what do they Involve?

Alternative holidays have the potential to meet many of the requirements of the good tourist. Many of these activities have, of course, been popular for a long time, while others are quite new; and many constitute the second holiday for some people. Nevertheless, the desire to do something different has provided a sufficient market and many small companies now cater to this growing demand and are often more aware of environmental issues than larger companies.

Two alternative options, which require more vetting than others, are personal development holidays and naturist holidays. Both of these are definitely 'off-beat', but they are also an illustration of how alternative need not necessarily mean 'environmentally friendly'. Apart from being able to strip off completely, naturist holidays are often really no different from a conventional beach holiday. Certainly they allow a freedom which is otherwise unattainable, but this does not exclude the potential for inflicting environmental or cultural damage. The responsibility for avoiding this lies with both the tour operator and the tourist.

Personal development holidays are slightly different. The success of these usually depends on a group of people working together at an activity which is designed to further self-knowledge. It can be reasonably argued that the basic principle of this is 'environmentally friendly'. A group 'looking at some of the barriers we build up towards love through a process of sharing and counselling' (see Cortijo Romero p.186) will hopefully learn about people the world over in learning about themselves.

The idea of being educated while on holiday is a common

factor to many alternative operators. The difference between them simply depends on what you are learning about. The New Experimental College in Denmark illustrates this: visitors from all over the world join with adult students from Denmark to help them study, and learn along the way. Courses are in English and there is no difficulty in becoming part of an international 'team'. Another company, appropriately named Study Tour Services, is run single-handedly as a specialist operator to Australia. The director stresses that, 'We are not concerned with the conventional beach holiday, but providing an intelligent framework to understand the country, including conservation and "green" issues.'

Attention should be drawn to another company much closer to home and perhaps more financially accessible than some of its exotic equivalents. Country Village Weekend Breaks is a 'community-based co-operative marketing initiative', based in the Marches counties on the border of England and Wales. Their aim is, 'to supply promotion at national and international level for the local providers of Bed & Breakfast, meals, guided walks and entertainment.' The organisation of this scheme comes from grass-root level with the onus for making it work placed on individuals and local communities interested in participating. For visiting foreigners this is the ideal chance to 'meet the Brits' and for the Brits themselves it is an opportunity to learn something about their own country. The whole scheme revolves around the idea that experience furthers knowledge and knowledge leads to understanding; in this instance of what the British countryside is about and how best to conserve it.

Up-market homestays are offered by companies such as Esther Eder's At Home in England and Scotland and Wolsey Lodges. Holidays with the British gentry offer homestay enthusiasts an alternative perspective on British life. Hosts are usually Britons of distinction, often with military, academic or political backgrounds and the emphasis of the holiday is on intellectual discussion with like-minded foreigners. In order to achieve this clients often have to fill out extensive application forms which ask for details of background and interests. In this way hosts are matched to guests.

Education is the idea behind tours run by Goodwill Holidays, who offer their 'Meet the Russians' programme to Westerners interested in seeing the 'real' USSR. This company is unique as the only one to offer this type of experience to travellers from Britain. Getting to grips with Russian culture can be difficult as a tourist but by meeting the Russians themselves, being invited into their homes and attending organised lectures, few visitors can fail to acquire a better insight than a conventional holiday would allow. The idea is to come away from Russia having established personal contacts and knowing that the stereotyped images usually portrayed will never seem authentic again. Although recreation does have a role to play in such a process, holidays are not exclusively leisure-based in the way that others are.

Smokestack Tourism

On the increase all over the UK is the promotion of tourism within city boundaries. 'Smokestack tourism', as it is often called, seeks to use to maximum effect facilities, resources, land and buildings within a city, often exploring the industrial and environmental heritage of cities developed during the Industrial Revolution. Sometimes this involves cleaning up particularly run-down areas. Frequently projects are geared towards educating people about how to make the most of their lives in the city. Organic farming and solar power are just two of the many issues which smokestack tourism promotes in an effort to highlight, amongst other things, the potential, possibilities and advantages of sustainable technologies. The format for taking part in one of these projects comes in two forms: either as a day-out, or as an alternative holiday when more extended periods of time can be devoted to learning new skills.

Practicalities

The practicalities of alternative holidays are really the same as those of any other holiday. The main difference lies in the fact that because for many people this will probably be the first time they've tried something different, extra care should be taken when checking a company's credentials and looking into what they offer. Cycling holidays in particular should be investigated thoroughly. In effect, these are 'soft adventure/activity' holidays and, although not as strenuous as mountain climbing, they do still demand a certain level of physical fitness. Make sure that you don't overstretch yourself simply because the brochure makes it all look easy and straightforward. It goes without saying that cycling holidays (depending on where you do them) can be one of the 'greenest' options available, just as long as people don't sink into self-satisfied complacency because they've managed to leave the car at home. Pedal-power itself may not harm the environment but, as on any holiday, tourist behaviour can still cause pollution.

Alternative Holidays Available

This list is by no means exhaustive. It offers some insight into those companies which, we believe, offer genuine alternative holidays, geared specifically to meeting your demands and meeting the principles of sustainable tourism. There will also be a few companies who, until now, have received little publicity. These are often the companies who have been brave enough to offer something really alternative or new but haven't the funds required for wide-reaching publicity campaigns. Their lack of money is not a reflection of the quality of their holidays but simply an indication that they truly are 'alternative' and have battled their way against the majority, managing to survive despite what are sometimes pretty unfair odds.

Special Activity, Hobby and Pastime

ALTERNATIVE TRAVEL GROUP.
Gentle walking in a group or independently is the idea behind this company's tours, which run throughout Europe, Turkey and India. Walks are continuous (not in circles) and researched fully to offer, 'the best and most picturesque views, the most interesting paths and the rarest flowers.' Luggage is transported separately and days are left free for sightseeing and relaxing.

Helpful brochure and well-constructed itineraries. Prices range from £745 to £1,500 for nine to fourteen days.

Further information from: 69–71 Banbury Road, Oxford OX2 6PE. Tel. 0865 251195.

ASTURIAS WALKING LTD.
Asturias Walking Ltd offer walking holidays in the Picos, one of Spain's most outstanding national parks, in the north of the country. It is significant and indicative of the company's policies that they start their brochure with the following introduction: 'We may thank Asturian ecologists for their efforts in keeping Spain's most beautiful mountains enchanting and unspoilt.' Tourism is stringently monitored and large developers have been kept out. First priority has been given to the preservation of the environment, wildlife, local culture and architecture. These efforts were rewarded in 1985 when Asturias won the European Preservation of Nature Award. Twenty walks were offered throughout 1991, with prices ranging from £284 to £806 for seven or fourteen nights.

Further information from: 6th Floor, Beacon Tower, Fishponds Road, Fishponds, Bristol, BS16 3HQ. Tel. 0272 584488.

BALKANTOURIST.
Balkantourist is Bulgaria's largest state tourist organisation and has been handling international tourism in Bulgaria for over forty years. Their holidays fall into three categories since as well as being a country specialist they also offer multi-pastime holidays and study tours. A significant section is devoted to special interest tours, covering culture, art, education, music festivals, national festivals, folklore and crafts, nature, sports, agricultural and industrial tours and incentive tours.

Further information from: c/o Bulgarian National Tourist Office, 18 Princes Street, London, W1R 7RG. Tel. 071 499 6988, or Balkantourist, 1 Vitosha Boulevard, Sofia 1000, Bulgaria.

CYCLISTS' TOURING CLUB.
Britain's national cyclists' association produce tours throughout the UK and abroad. Average price eight days in Wales:

£95. Brochure also gives information on cycle holidays run by independent companies.

Further information from: CTC HQ, 69 Meadrow, Godalming, Surrey GU7 3HS. Tel. 0483 417217.

DALLAWAY'S VENTURES.

Holidays to Czechoslovakia and Russia where the client helps to 'restore the state of the environment; become involved within the community and learn of the rich history and culture.' Prices £857 by air for eighteen-day trip to Czechoslovakia to restore hunting lodge in Harrachov, Northern Bohemia; £734 by bus.

Further information from: 80 Stuart Road, Wimbledon Park, London, SW19 8DH. Tel. 081 946 6295.

DETOURS.

An all-round introduction to countries in Eastern Europe, Africa, Asia, the Caribbean, and Latin America. As well as covering 'traditional' cultural elements such as the arts and architecture, participants also learn about people's everyday lives through home visits and home stays. All the tours contribute financially to conservation and development projects, and in some cases there is an opportunity to work on a project too. The holidays usually include a visit to a national park – and there is plenty of opportunity for relaxation! Travel is always in small groups of a mixed age range. The trips cost £700–£1,800 for two to three weeks.

For further information: 80 Stuart Road, Wimbledon Park, London, SW19 8DH. Tel. 081 946 6295.

GREEN HORIZONS.

Green Horizons organise personal introductions, 'off the beaten track', to the Marches country of western Herefordshire and eastern mid-Wales, including the Radnor Hills and the Black Mountains. Guided walks and mini-bus tours are organised according to personal interest. The guide is a landscape ecologist with professional experience of landscape and nature conservation, and with special knowledge of local history and local literary figures. He has special dispensations to visit local estates, woods and parks. Tours include drove roads, ridgetop tracks, Offa's Dyke Path, Kilvert country, botanic gardens, ancient woods, deer parks, hill-forts, castles, mottes and churches. Prices are £25 for an evening, £30 for a half day and £60 for a full day.

Further information from: Roger Crawley, Little Pentre, Pentre Lane, Bredwardine, Hereford, England, HR3 6BY. Tel. 09817 467.

HEADWATER HOLIDAYS.

The difference between Headwater's walking holidays and those which are listed under Activity/Adventure (see Chapter 11) is the independence allowed by their flexible system and the fact that walkers make their own way *at their own pace*. This is certainly alternative compared to the adventurous 'climb a mountain' type of walk offered by many other companies. Headwater will organise more 'strenuous' walks if so desired but tend to let the clients determine the level of fitness required. Basing themselves in little-visited areas of France, Headwater try to integrate clients into the local culture as much as possible, making use of local facilities and inviting walkers to really absorb their surroundings without intruding unnecessarily into the established way of life. New for '92 are the cycling and walking holidays in Burgundy, and multi-activity centres in the South of France and Ardèche. Prices range from £300 to £500 for six to ten nights.

Further information from: 146 London Road, Northwich, Cheshire, CW9 5HH. Tel. 0606 48699.

HF HOLIDAYS.

HF Holidays is a specialist firm dealing in walking holidays, often through national parks. It is a non-profit making organisation founded over seventy years ago, offering a range of wildlife and special interest holidays in Austria, Spain, France and Switzerland, set mainly in areas of outstanding natural beauty. Fourteen days in Switzerland costs about £500.

Further information from: Imperial House, Edgware Road, Colindale, London, NW9 5AL. Tel. 081 905 9556.

HOSKING TOURS LTD.

A small company dedicated to natural history photography. Tours to best worldwide locations for bird and mammal photography. Average price for fourteen days India 'Land of the Tiger' tour £2,500.

Further information from: Pages Green House, Wetheringsett, Stowmarket, Suffolk, IP14 5QA. Tel. 0728 861113 Fax. 0728 860222.

INNTRAVEL.

Cycling is just one of the various pastimes offered by Inntravel in their European brochures. The company also deals with short-breaks, painting, walking, fishing, sailing, wine-tasting, horse-riding and skiing (the last is very much cross-country, due to the less harmful impact on the environment). Inntravel are conscious of the potential damage their tours could cause and their walking holidays, for example, 'choose areas and footpaths which do not threaten natural habitats' and try to ensure that 'the numbers of walkers will not have harmful effects upon the environment.' On wine-tasting excursions local rather than major names are used and throughout care is taken never to 'anglicise' products. As far as Inntravel is concerned, 'when the local culture is submerged in the tourism around it, that destination ceases to be of interest.' Clients travelling with Inntravel usually do so independently. The company's rather complex brochure does have the advantage that travellers can build up their own itinerary, choosing what to do, when and where. A basic five-night cycling holiday around Vosges and Alsace costs between £255 and £420.

Further information from: The Old Station, Helmsley, York, YO6 5BZ. Tel. 0439 71111.

INSIGHT TRAVEL.

Host-based holidays in the Ashanti region of Ghana, sold individually, so totally flexible. Video showing holiday available by post for £7 gives excellent insight into life in Ghana and what can be achieved on a host-based holiday. Average price fourteen nights £490, not inc. flight.

Further information from: 6 Norton Road, Garstang, Preston, Lancs. PR3 1JY. Tel. 0995 606095 Fax. 0995 602124.

Alternative Holidays 183

MAZORCA RESEARCH DEVELOPMENT EXCHANGE.

Cultural tours to Cuba aimed at the independent responsible traveller interested to learn more of other cultures. Attending one of the Cuban cultural events is often the highlight, but several add-ons available. Weekends in Mexico, or Bahamas, etc. Sample tour Folk Cuba, based in Havana, fifteen days £750+.

Further information from Hackney Business Centre, Studio 8, 277 Mare Street, London, E8 1EB. Tel. 081 533 5432 Fax. 081 533 6996.

OLAU LINE (UK) LTD.

As part of their general brochure, the ferry company Olau Line are offering cycling holidays covering the whole of Holland. Accommodation is offered in hotels belonging to the Postiljon group, so money is at least staying in the local economy. If considering a tour with Olau the emphasis while cycling is very much on the client adhering to his own 'environmental rules'. The six-day tour costs £231 per person, for two people travelling together. This price becomes cheaper with the more people involved but the limit is six people per car.

Further information from: 104 Anchor Lane, Sheerness, Kent, ME12 1SN. Tel. 0795 666666.

PENG TRAVEL.

Naturism is a 'family affair' according to Peng Travel. The company offers naturist holidays to Crete, Yugoslavia, Turkey, Spain, Florida and the French Caribbean. The enjoyment of these holidays stems from, amongst other things, 'the unique sense of freedom that nakedness brings.' A seven-night holiday at Costa Natura costs between £400 and £530.

Further information from: 86 Station Road, Gidea Park, Romford, Essex, RM2 6DB. Tel. 04024 71832.

SUSI MADRON'S CYCLING FOR SOFTIES.

The name of this company reflects the spirit in which the directors approach their holidays. Established now for over ten years, the company has received much media attention for their work, not least of which is the recognition that they are an 'environmentally friendly' tour operator well suited to the independent traveller's needs. Bikes have been specially designed for the company and holidays can be as relaxing or strenuous as you like. Prices vary but a typical example is the ten-night 'Gentle Tourer' around Provence, costing from £590 to £640.

Further information from: Lloyds House, 22 Lloyd Street, Manchester, M2 5WA. Tel. 061 834 6800.

TRISHULA TRAVEL FOR A NEW ERA.

Staying in the homes of locals, tours to various places in South America celebrating the folklore of the country, and led by a professional guide. Sample price £1,650 for eighteen days in Venezuela.

Further information from: 44 Alconbury Road, London, E5 8RH. Tel. 081 806 4916.

Education

ADULT EDUCATION STUDY TOURS.

Run in conjunction with experts from five British universities, this company's tours are, 'holidays with the added pleasure of learning more about a culture, society or environment.' Trips are categorised into two areas: 'Cultural Study Tours' and 'Contemporary Life Study Tours'. They are offered throughout the world. Adult Education offer what is probably one of the most diverse and varied programmes to be found, at prices which give you more for your money than many of the mainstream operators. A typical price is £645 for the Italian idyll trip. For '92 a new range of tours to the USSR is being developed.

Further information from: 49 The Mall, Faversham, Kent, ME13 8JN. Tel. 0795 539744.

ASSOCIATION FOR CULTURAL EXCHANGE (ACE) STUDY TOURS.

ACE are now in their thirty-fourth year of service, offering trips which reach 'all corners of the globe': from Orkney to the Galapagos Islands, North and South America, Namibia to Burma, the Isle of Mull to New Zealand. Their aim is to 'provide a series of stimulating programmes with a cultural theme'. Included are trips specifically on architecture, art history, archaeology, music, drama, ecology and wildlife. Prices vary considerably, but the following are two typical examples: a week's tour of Italian Lakes, Villas and Gardens at £535, while a fourteen-day tour along the 'Road to Samarqand' costs £1,455.

Further information from: ACE Study Tours, Babraham, Cambridge, CB2 4AP. Tel. 0223 835055 Fax. 0223 837394.

CAMBRIDGE STUDY HOLIDAYS.

Courses run at Cambridge University are held throughout the summer, for two to six weeks. Day schools also available. Subjects are all centred on Britain, covering everything from British political thought to the history of English painting. Accommodation is provided in temporarily-vacated student lodgings, often in one of the university's historic buildings. A typical two-week course costs roughly £500, including all tuition, two meals per day and accommodation.

Further information from: The Director, Board of Extra-Mural Studies, Madingley Hall, Madingley, Cambridge, CB3 8AQ. Tel. 0954 210636.

NEW EXPERIMENTAL COLLEGE.

Denmark's New Experimental College (NEC) is a Danish Folk High School which seeks to bring together people from all over the world and widely different backgrounds to study together a subject of their choice at the same time as learning about Denmark. The college offers applicants the choice of study at a more conventional school, or in the more radical Experimental College itself, where students set their own targets, discipline their own work and report every Saturday on what they have achieved during the past week. Integration, social interaction, self-realisation and self-analysis are central to the college's aims.

Further information from: Skyum Bjerge, Thy, 7752 Snedsted, Denmark. Tel. 010 45 97 936234.

OXFORD STUDY HOLIDAYS.

Study Holidays at Oxford University are available from July to mid-August as either a three- or six-week course. Accommodation is provided in what are usually student 'halls' in various colleges (some dating from the fourteenth century). Many subjects are on offer and an average price is £250 per week.

Further information from: Summer School Secretary, Department for Continuing Education, Rewley House, 1 Wellington Square, Oxford, England, OX1 2JA. Tel. 0865 270360.

STUDY TOUR SERVICES.
The director of Study Tour Services has been arranging study tours, summer schools and conferences for over thirty years. Although his trips are now mainly to Australia, he will also arrange and lead tours to China, Russia, the US and other locations on request. He regards travel as an extension of education and orientates many of his tours around the problems of environmental conservation, working in cooperation with relevant conservation bodies. A three-week all-inclusive tour costs about £2,000.

Further information from: 39 Burgess Road, Bassett, Southampton, SO1 7AP. Tel. 0703 790691.

Spiritual/Holistic

CORTIJO ROMERO.
Cortijo Romero has been running for five years as a 'centre for spiritual and personal development'. Its ethos lies in the belief that 'love is the source of transformation in our lives' and that 'we have no one creed or discipline'. Periods of what can be fairly intensive 'learning' are mixed with free time to explore the local countryside of southern Spain. Meals are all vegetarian with an emphasis on grains, pulses and locally grown fruit and vegetables. Cortijo Romero state, 'The courses give you the opportunity to discover more of yourself . . . through movement . . . song . . . group discussion, sometimes encountering pain, sometimes laughter, to find your natural joy and inner peace.' Prices range from £230 to £630 for seven to fourteen days, not including flights.

Further information from: Susan Lever, 4 Malden Road, London, W3 6SU. Tel. 081 993 4898.

SKYROS CENTRE.

The Skyros Centre offers holistic and personal development holidays on the Greek island of Skyros from April to October. Widely praised in the press and media, this company runs three centres: Atsitsa, 'a unique adventure in health and fitness'; the Skyros Centre, 'in search of a way of living that can help us renew our connections with other people, with our environment, with our sense of purpose and ultimately with our deepest selves'; and the Skyros Institute, 'training courses in personal development and holistic health within a theoretical framework which assumes that health and well-being relate to the development of one's physical, emotional, social, intellectual, and spiritual potentials.' All three centres are on Skyros. Prices range from £445 to £580 for almost two weeks.

Further information from: 92 Prince of Wales Road, London, NW5 2NE. Tel. 071 431 0867.

WESTERN BUDDHIST ORDER.

Organised through the London Buddhist Centre, the Western Buddhist Order offer a number of retreats throughout the year. In terms of personal development and raising awareness of self and others, they are highly successful and expertly instructed. Retreats are held in Norfolk, Wales, Shropshire, the mountains of south-east Spain and in India and are both mixed and single-sex. The London Buddhist Centre also offers day/weekend courses and classes in Buddhism in general, meditation, yoga, massage, shiatsu, t'ai chi, homoeopathy, acupuncture and aromatherapy, to mention a few. The cost of retreats varies. Some are based on £25 a night while others have a set price of £60 and £90. Low-income retreats are available for £8 per night for the unemployed, students or those simply on a low wage.

Further information from: London Buddhist Centre, 51 Roman Road, Bethnal Green, London, E2 0HU. Tel. 081 981 1225.

Alternative Technology, Living and Travel

ALTERNATIV BUS REISEN INC (ALTERNATIVE BUS TRAVEL).

Alternativ Bus Reisen was started by a group of Hamburg friends who had chartered a bus for a European holiday. Their tours, ranging from skiing trips in Europe and the US to Caribbean schooner sailing, are aimed at serious travellers interested in learning about the places visited. On average, hotels are used on a third of the nights spent travelling, with the bus itself and tents providing 'accommodation' at all other times. Although tours are accompanied by a 'tour leader', there are no formal talks and it is up to travellers to track down information for themselves. The emphasis throughout trips is very much on low-budget travel.

Further information from: 136 East 57th Street, Suite 1700, New York, NY 10022, USA. Tel: 212 421–2212.

BRIDGEWATER BOATS.

With a maximum speed of 4 mph barging is bound to be relaxing, slow and peaceful. A family-run concern, this company offers eleven attractive old-style barges for holidays round the south-east of England, including the Chiltern Hills region. Boats sleep four to eight people and are well equipped. Beginning at Berkhamsted you could take in Bulbourne and the nature reserve at Tring, then on to Marsworth and Aylesbury. If you've a longer spell you could steer a course right up to London. Approx. cost for six-berth, £650 per week.

Further information from: Castle Wharf, Berkhamsted, Herts. Tel. 0442 863615.

YOUTH HOSTELS ASSOCIATION.

The Youth Hostels Association (YHA) offers low-budget accommodation throughout the world to anyone of any age in need of a bed for the night. Increasingly popular with the 'older generation' as well as with back-packing students, hostels provide the perfect opportunity to meet fellow travellers from all over the world, living and working (for an hour each day) together and taking part in any of the many activities which are usually on offer. Average price for a bed for the night is £6 to £8. The YHA and SYH also organise activity courses throughout the year.

Further information from: Trevelyan House, 8 St Stephen's Hill, St Albans, Hertfordshire, AL1 2DY. Tel. 0727 55215. In Scotland: Scottish Youth Hostels, 7 Glebe Crescent, Stirling, FK8 2JA. Tel. 0786 51181.

Community Living, Homestays and Homeswaps

BRITISH UNIVERSITIES ACCOMMODATION CONSORTIUM LTD.
On campus accommodation outside term times at reasonable prices. Summer schools also available.
Further information from: PO Box 705, University Park, Nottingham, NG7 2RD. Tel. 0602 504571 Fax. 0602 422505.

COLBY INTERNATIONAL.
Colby International is the UK agent for anyone wishing to make Bed & Breakfast bookings in cities throughout the US. Prices start at roughly £30 a night, bed & breakfast for a double room.
Further information from: 139 Round Hey, Liverpool, L28 1RG. Tel. 051 220 5848.

COUNTRY HOMES AND CASTLES.
Country Homes and Castles offer clients the chance to stay in some of Britain's most elegant and grand private homes, as guests of the owners. They also organise bookings for a selection of British Country House Hotels, renowned for their old-fashioned grace and service. Properties are spread right across the country, from an eight-bedroomed Georgian listed building on the banks of Loch Ness to a nine-bedroomed house dating from the fourteenth century in Hampshire, reputed to be where William Thackeray wrote most of *Vanity Fair*. The company also offer a selection of cottages situated in the grounds of large estates and will organise extras such as tickets for the Chelsea Flower Show, Wimbledon, or the Edinburgh Festival, as well as reservations at restaurants, conferences and incentives and vintage Rolls Royce Tours.
Further information from: 118 Cromwell Road, London, SW7 4ET. Tel. 071 370 4445 Fax. 071 370 4449.

COUNTRY VILLAGE WEEKEND BREAKS.
Country Village Weekend Breaks aim to reveal the unique nature of English villages by inviting visitors to stay with one of the locals in their community and to see that community through local eyes. As well as making a direct contribution to the local economy the

organisation advocates '"Promotion with protection" of the social and physical environment in areas where any tourism development can be viewed with suspicion or alarm.' Villages are situated on the border of England and Wales and in the Lakelands, where locals are trying to attract tourists away from the saturated areas of Bowness and Windermere. Family History weekend also available from 1992. Average cost for a weekend break is £90.

Further information from: The Cruck House, Eardisley, Herefordshire, HR3 6PQ. Tel. 054 46 529.

DUBLIN SCHOOL OF ENGLISH LTD.
The Dublin School of English was originally set up twenty years ago as a centre for teaching English as a foreign language. Today the school organises homestays for anyone interested in visiting Northern and the Republic of Ireland. If staying in Dublin clients are also entitled to join in the school activities.

Further information from: 10–12 Westmoreland Street, Dublin 2, Ireland. Tel. 773322.

EN FAMILLE OVERSEAS.
A small company arranging for paying guests to slot into families in several European countries, though specialising in Host Families in France. They also run language courses, in suitable centres in France, as well as House parties and Young People's Holidays. Costs vary depending on family and location, but half board per adult is approx. £130 a week. Joining fee £35. Excellent pre-departure literature provided and guests report a genuine integration with the locals, not a stage-managed experience.

Further information from: The Old Stables, 60b Maltravers St, Arundel, W. Sussex. Tel. 0903 883266 Fax. 0903 883582.

EXPERIMENT IN INTERNATIONAL LIVING.
A non-profit making, non-political and non-religious organisation, leading in international education and exchange since 1932. Promotes international homestay programmes which will foster mutual friendship and understanding of different nations. Operates in thirty countries and also offers outbound programmes: an au pair homestay programme, Volunteer Work Programmes,

Youth Exchanges (including Young Workers, Disabled and Disadvantaged) and International Student of English Language campus courses. Ages 16+, one to four week courses. Average price one week homestay + one week Volunteer Work Programme £300.

Further information from: Otesaga, West Malvern Road, Malvern, Worcs. WR14 4EN. Tel. 0684 562577.

GOODWILL HOLIDAYS LTD.

See what life in the USSR is really like with 'Meet the Russians'. The stated aim of this company is, 'to provide an opportunity to gain an insight into the culture, society and politics of the USSR, to discuss the recent exciting changes with experts and lay people alike and perhaps to return home with a more sympathetic understanding of life in the Soviet Union.' Special interest tours can be arranged for those from virtually any walk of life, as well as tours to significant Russian cities. All tours are accompanied by a company leader as well as an experienced Soviet interpreter-guide. Prices well below Intourist equivalents (e.g. £520 for a week in Moscow, full-board).

Further information from: Manor Chambers, School Lane, Welwyn, Herts, AL6 9EB. Tel. 043871 6421 Fax. 043871 7262.

THE GREEN THEME HOME EXCHANGE HOLIDAY SERVICE.

Offers a personalised service and tries to promote home exchange because it is 'a very environmentally friendly way of holidaying without putting undue stress on the host country'. Annual membership £10. Also offers Short Break Exchanges and UK Exchanges for an additional £15 – an excellent way to try out home-swop. Also offer 'Hospitality Exchange' whereby you act as hosts to visitors in your home on the agreement they will reciprocate in future.

Further information from: Little Rylands Farm, Redmoor, Bodmin, Cornwall, PL30 5AR. Tel. & Fax. 0208 873123.

HIGHER EDUCATION ACCOMMODATION CONSORTIUM LTD.

As BUAC, accommodation and activity learning holidays outside term times in Britain's polytechnics and colleges.

Further information from: 36 Collegiate Crescent, Sheffield S10 2BP. Tel. 0742 683759 Fax. 0742 661203.

HOMESITTERS LTD.
An agency offering vetted, reliable people to look after your home while you're away. To become a homesitter or have your home and garden 'sat', further information from: Buckland Wharf, Buckland, Aylesbury, Bucks, HP22 5LQ. Tel. 0296 630730.

INDRALOKA HOME STAYS.
Mrs Moerdiyono set up Indraloka Home Stays in 1970, following the success of inviting travellers to stay in her own home. Homestays are available in Djakarta, Yogyakarta, Bandung, Surabaya, Malang and every other major Javanese location. Particularly popular with single women, as well as travelling businessmen, Indraloka places clients in English-speaking homes, usually in large rooms with ceiling fans and offers the best of Indonesian home-cooking. Dinner, bed and breakfast approx. US $35 per couple.

Further information from: Mrs B Moeridiyono, 14 Cik Di Tiro, Yogyakarta 55223, Indonesia. Tel. 010 62 274 3614.

INTERCULTURAL EDUCATIONAL PROGRAMMES.
This is a registered charity, offering educational exchanges throughout Europe, North America, South America, Asia and Australia. Students, usually aged sixteen to eighteen, spend up to one year living with a host family and attending school, becoming fully integrated into the national way of life. The organisation works on the belief that 'It's fun being a foreigner', teaching youngsters the benefit of travel and the validity of learning about different cultures. The full fee for the year is £3,450, but there are grants of up to £3,000 available. Anyone interested in becoming a host family should also contact the National Office.

Further information from: IEP, Ground Floor Suite, Arden House, Main St, Bingley, W. Yorks, BD16 2NB. Tel. 0274 560677 Fax. 0274 567675.

INTERVAC INTERNATIONAL HOME EXCHANGE.
Home-swapping in forty-five countries from the leading UK specialist. New in '91 were Eastern European destinations – an excellent way to integrate with the locals there! £37 annual membership for the directories and help, then it's up to you

to start corresponding with people advertising for a UK swop. Youth Exchange programme also operated.

Further information from: 16 Siddals Lane, Allestree, Derby DE3 2DY. Tel. & Fax. 0332 558931.

NATIONAL TRUST HOLIDAY COTTAGES.

Everything from family-sized farmhouses to tiny bijoux cottages, old school houses, coastguards' houses, to historic mills, the National Trust offers properties for rent, but early booking is required. Brochure 50p and further information from: PO Box 101, Melksham, Wilts, SN12 8EA. Tel. 0208 73880.

SERVAS.

SERVAS is one of the main agencies involved in linking its members with each other for the purpose of hospitality-based visits. There are roughly 10,000 members worldwide, all interested in hosting visitors from abroad in return for a similar arrangement when they, the hosts, travel themselves. Annual subscription £20.

Further information from: Hazel Barham, SERVAS Traveller Secretary, 41 Pendre, Brecon, Powys, Wales, LD3 9EA. Or Ges Souttar, SERVAS, Bankside Cottage, Welton-le-Wold, Lincs. LN11 0QT. Tel. 0507 602512 Fax. 0507 600893.

VISIT BRITAIN.

Visit Britain offer British homestays in both 'standard' and 'upmarket' homes. Staying with British families, guests have the chance to learn about the traditions and culture of Britain, often joining in with the host's leisure and social activities. Average price for a week's homestay, plus two meals a day is roughly £110.

Further information from: Geraldine Wheeler, 11 The Croft, Hastings, East Sussex, TN34 3HH. Tel. 0424 431438 Fax. 0424 718456

WOMEN WELCOME WOMEN.

Network of women fostering international friendship by enabling those of different countries to visit one another. Runs on donations. All types of women: young, old, housewives, professionals, rich, poor invited. You may be a hostess, a traveller, or both. Individual visits organised by writing to hostess.

Further information from: 8a Chestnut Avenue, High Wycombe, Bucks, HP11 1DJ. Tel. 0494 439481.

The most straightforward way to arrange an overseas home swap is to advertise, for a charge, in one of the home swap directories listed below.
Home Base Holidays: 7 Park Avenue, London, N13 5PG. Tel. 081 886 8752. Home Interchange: 8 Hillside, Farningham, Kent, DA4 0DD. Tel. 0322 864 527. Homelink International: 84 Lees Gardens, Maidenhead, Berks 5L6 4NT. Tel. 0628 31951. Intervac: 6 Siddals Lane, Allestree, Derby, DE3 2DY. Tel. 0332 558 931, Britain's biggest directory. Worldwide Home Exchange Club: 13 Knightsbridge Green, London, SW1X 7QL. Tel. 071 589 6055. Christian House Exchange Fellowship (CHEF): Peter Worsley, National Organiser, Karakorum, Sunnyfield Lane, Uphatherley, Cheltenham, Gloucestershire, GL51 6JE. Tel. 0242 521 886. For church members.

Unusual Holidays/Unusual Places

ANDREW BROCK TRAVEL.
Established in 1981 and now operating under five different names, Andrew Brock Travel (also Martin Randall Travel) offers art and architecture tours to Jordan, North Yemen and Thailand, as well as throughout most of Europe. Highly regarded and using local facilities where possible, they are worth considering by serious art lovers. Tours are accompanied by a professional lecturer and are carefully planned to allow for maximum 'viewing' time. A seven-night tour to Italy costs about £825.

Further information from: 10 Barley Mow Passage, London, W4 4PH. Tel. 081 994 6477 Fax. 081 724 1066

BRITISH WATERWAYS BOARD.
The British Waterways Board have been involved with smokestack tourism for several years. Working in conurbations such as Bradford, Birmingham, Manchester and Liverpool, they offer

insight into the way city tourism can help redevelop inner-city areas, especially through pioneering canal boat holidays.

Further information from: Customer Services Dept, Grey Caine Road, Watford, WD2 4JR. Tel. 0923 226422.

INSCAPE FINE ART TOURS.

Founded in 1986, Inscape Fine Art Tours offers excellent tours to travellers, tourists and anyone interested in art, architecture and fine art. Their stated aim is, 'to provide standards of teaching and administrative excellence second to none.' One-day and residential courses are offered, mostly in London or around the Cotswolds, as well as European study tours to France, Germany, Austria, Italy, Holland and Ireland. Day tours cost around £20; weekend tours cost about £375, all-inclusive and European tours range from £700 to £900. Tours are highly recommended.

Further information from: Austins Farm, High Street, Stonesfield, Oxfordshire, OX7 2PU. Tel. 0993 891 726.

MAJOR AND MRS HOLT'S BATTLEFIELD TOURS.

Established over fifteen years ago, Major and Mrs Holt's Battlefield Tours are now run in association with the Imperial War Museum. The company is one of the leading specialists in battlefield tours, covering almost every area of note throughout the world, from the Somme to Gallipoli to the Falklands. Their thought-provoking tours bring home the stark realities and absurdities of war in a useful way. Tours are accompanied by guest lecturers, often ex-Service military historians and much use is made of literature, music, newspapers and videos. Prices range from £155 to £1,700 for two to fourteen days.

Further information from: The Golden Key Building, 15 Market Street, Sandwich, CT13 9DA. Tel. 0304 612248.

NATIONAL CENTRE FOR ORGANIC GARDENING.

Open throughout the year, the National Centre for Organic Gardening was established in 1986 by the Henry Doubleday Research Association. Visitors to the twenty-two-acre site are offered insight into all forms of alternative/organic gardening, including such topics as biological pest control, 'no-dig' gardening and the use of compost fertilisers.

Further information from: Ryton-on-Dunsmore, Coventry, CV8 3LG. Tel. 0203 303517.

VEGIVENTURES.

Offer UK and European holidays for vegetarians. Small groups, low impact, use of public transport and organically grown food. Sample price: one week in Highlands of Scotland on multi-activity break £255; fourteen nights in Portugal £589. Further information from: 17 Lilian Road, Burnham-on-Crouch, Essex, CM0 8DS. Tel. 0621 784235.

FARMING/CROFTING HOLIDAYS.

Many of these now exist. For Highland crofts, contact the Highlands and Islands Development Board and ask for their brochure 'Self-Catering and Special Interests'. Tel. 0349 65000. For English farms contact the English Tourist Board on 081 846 9000. Most foreign countries have London-based tourist boards who will send out literature on their country's farming holidays. One tour operator who offer well-priced farms in Norway is Color Line. Tel. 091 296 1313.

Chapter 11

Activity/Adventure Holidays

The market in adventure and activity holidays has more than trebled in the past eighteen months. Outdoor pursuits and the challenge of trying a new sport, exploring new terrain, or enduring the dare-devil thrill of an activity such as hang-gliding, have all risen in popularity in conjunction with the rise of the somewhat misplaced belief that 'being outside is being green'. This, of course, is not necessarily true, but it is a belief that some tour operators have been quick to capitalise on, passing off holidays as 'environmentally friendly' when what they really mean is, at best, health conscious and physically demanding; at worst, environmentally destructive and physically hazardous. This chapter attempts to bring some sort of order to a very large category and to indicate which operators are genuinely trying to avoid misuse of countryside, rather than hiding under a label which they believe will earn them popularity.

What is an Adventure/Activity Holiday?

Adventure and activity holidays take many different forms. It is often difficult to decide what should be classed as adventure/activity and what would be better labelled simply as sport, walking, driving, or even very generally as 'special interest' holidays. Companies vary in size, from the well established ExplorAsia, working as the UK agent for the international Tiger Mountain group of companies, to smaller concerns such as Dick Phillips, who, although having operated now for more than thirty years, continues to specialise solely in tours of Iceland, building on experience and educating clients about their surroundings. The range of approximately seventy five companies is really quite staggering and travellers can enjoy adventure holidays all over the world.

The common factor of many of them, apart from sports holidays for golf or tennis, is the contact they allows with wilderness. If undertaken properly this can be very rewarding for the individual, contrasting completely with the frenetic, urbanised lifestyles most of us lead. The experience can also educate people to appreciate wild places and wildlife and can lead to an understanding of the importance of conservation. The key to achieving 'good tourist' status, however, lies in the ability to distinguish not only between the green holiday and those that are simply outdoor but also to assess the approach and sensitivity of the companies which offer them.

The attraction behind an adventure holiday is the desire to try something new and discover different places, or be pushed to one's limit, the driving force for many people who participate in such ventures. The companionship also helps to enhance the enjoyment of what, undertaken independently, might feel more like a solitary struggle than an active holiday. Many people have been persuading themselves that the public school 'stiff upper lip' attitude is a satisfying test of endurance, since it does us good to become hardened to the rigours of life and, besides, adversity strengthens the character . . . doesn't it?

An adventure holiday is not about putting yourself through hours of physical torture, suffering silently until your hands are numb with cold rather than admitting your own limits. It is about benefiting from the expert tuition of specialists in fields as diverse as GCSE fieldwork for schools in Morocco, to exploring crater lakes and volcanoes in South America. Communication with, rather than domination of, nature is a key factor: in testing oneself against nature, the outcome is often a new respect for forces previously overlooked or taken for granted. In one way or another an adventure/activity holiday should be an educational experience where personal effort is rewarded by the acquisition of a new skill, or of knowledge previously denied through lack of experience.

How Green are Adventure/Activity Holidays?

The irony of adventure holidays is that while they can be personally rewarding, it is that very regard for the self which, in its complete preoccupation with personal reward, can lack any sensitivity. The concept of being 'environmentally friendly' demands an awareness of others and, perhaps more importantly, of surroundings other than those immediately in front of you, if it is to be at all successful. That's not to say that adventure holidays, cannot be environmentally friendly, but simply that many tour operators fail to meet, or are unaware of, the criteria, which, in our view, are essential for a caring approach to the environment. It's not much good to anyone if having risen to the challenge of scaling a mountain, you then drop your non-biodegradable sandwich wrappers, descend and insult the locals. You may have acquired the know-how needed to climb Everest, but your presence could still give offence, damage culture and lose hard-earned respect.

The responsibility for maintaining a sensitive approach does not lie solely with the visitor. Any company offering what they term an 'adventure/activity' holiday should take adequate measures to ensure that clients are well educated about their destination, even if it is as close to home as the north of Scotland, and that they are aware of how to get the most out of their holiday without having a detrimental impact on the place they are visiting. In the view of the tour operator, the idea of a holiday is often to take and not to give, apart from the cash which you hand over as payment. You should be wary, for example, of a company which likes to tell you of the contribution it is making by using 'local transport' and 'accommodation likewise', but which also proudly declares that it uses its 'own trusted tour leaders . . . unlike other companies who'll fob you off with local guides'. Although it is sometimes preferable to have a British representative, it is seldom absolutely necessary. The wording of the above 'assurance' carries the implication that if you're 'fobbed off' with a local guide, you're getting a raw deal. This, of course, is not necessarily true.

Another aspect of adventure holidays is that contact with locals is often minimal, since, whether at home or abroad, it is the countryside and facilities which the tourist wishes to enjoy and personal contact often takes second place. It's easy and understandable for participants to acquire the blinkered view that their instructors and the skills they impart, are all that matter, regardless of what country they might be in. This is a problem which is encountered in the adventure/activity field more so than in others.

Skiing

Skiing now constitutes one of the most popular types of holiday on the market. It represents for many people one of the great dilemmas and conflicts between recreational enjoyment of the countryside (and the associated development of ski-lifts, restaurants, and other facilities with the economic benefits for rural areas desperate to attract investment) and the conservation of the fragile alpine and mountain areas where skiing inevitably takes place. The situation is particularly acute in the Alps where, in recent years, the issue has been raised by many concerned environmentalists. The same dilemma of recreation and development versus conservation has been vigorously debated in the Cairngorms in Scotland. The continual use of the same location and of the same runs at that location, together with pressure to expand the skiing areas has brought skiers and conservationists into conflict. Unfortunately there is no way of compensating for the damage done on the ski slopes and it is often permanent rather than temporary. The result is that the face of the landscape changes and eventually the whole ecosystem of a particular area is altered.

Regardless of measures taken by designers to incorporate ski complexes into the surrounding environment, the construction of ski facilities inevitably scars the landscape even before the skiers have arrived. Pylons supporting chair-lifts, new hotels and access roads can usually be seen from miles around. In the winter snow the visual effect becomes less harsh and the scene has become an accepted sight of winter sports. A useful exercise is to visit a ski resort in the summer and see the damage in a more realistic light. During the rest of the year there is little that can be done to compensate for the intrusion of complexes into the beauty of the surrounding countryside, let alone for the damage caused to wildlife during the initial construction period.

The question of responsibility towards the environment is one which skiing authorities sometimes seem reluctant to accept. It is understandable that fear of losing business might prevent them from making changes which could alter the whole nature of their sport, yet there are some who even seem unwilling to admit the fact that skiing can damage the environment. The problems involved with the promotion of skiing cover a variety of issues. In Scotland and the Alps demand has now exceeded supply and it is claimed that further development is required if the industry is to continue to function successfully. Conservationists who are worried by the thought of this might consider the implications of *not* developing further: land already under pressure will suffer more severely as greater numbers of skiers scramble to reach the slopes and habitats that have survived so far are likely to be irrevocably damaged.

In the Alps one of the main problems has arisen from the construction of access roads from widening footpaths. Traffic in this area – both vehicular and 'human' – has increased dramatically in the past thirty years. An estimated 100 million people visit the Alps each year and with them have come the problems of pollution and erosion. Destruction has been caused by deforestation and altering the use of traditional alpine land for the construction of dams, skiing facilities and hotels and by the dumping of waste which has polluted nearby lakes. Reports from the Swiss-based World Conservation Union and the International Centre for Alpine Environments in France conclude that demand for skiing facilities is growing by 5 per cent each year. Without controls, there will soon be little left to see of the Alps, apart from ravaged countryside lying as a grim reminder of the past.

Mountain Wilderness, an organisation established in 1987 and known as the 'Greenpeace of the Alps', has dedicated itself to raising awareness of these problems and to preventing further damage. Among their members can be found figures such as Reinhold Messner, Chris Bonnington, Lord Hunt and Sir Edmund Hilary, the honorary president. They have succeeded in preventing the building of constructions such as the ski-circuit of Mount Pelmo, ski-lifts on the Chaviere glacier in the Vanoise and a ski resort at Saleve. The interest is such that this initiative is being taken up in other alpine areas around the world.

Other projects underway include help from the Russians to

repopulate the Pyrenees with the once-prolific brown bear. Hunting of this animal was legal in France until 1962 and in Spain until 1975. Hunting combined with damage and intrusion from skiing has all but wiped them out and today there are believed to be only fifteen bears left in these areas. The situation is changing, however, thanks to the Russians who are sending about fifty bears to the Pyrenees in order to restock the population.

For anyone interested in skiing then, it might seem that there is little they can do to right the wrongs unless they never ski again. Alpine skiing is generally more destructive than 'lang-lauf' or cross-country skiing and individuals might consider changing to this type of activity. A large responsibility undoubtedly lies with tour operators and the skiing industry itself, which should be seeking alternative places to which to send their clients and limiting the numbers present in any one season. What the client can do is to ask operators whether any type of monitoring is carried out; whether the company takes any active role in researching environmental damage caused by skiing and, generally, what solution the company might offer by way of improving the situation. Such questions are hardly out of place and any responsible operator should be willing to answer your queries. After all, it's in the industry's own interest to ensure the conservation of the countryside they use and the more people who become involved in doing that, the easier the task should be. If your questions are continually avoided you can always seek alternative companies who display a more responsible attitude. Skiing is big business and the holidays are not cheap: within the industry it is also highly competitive and it wouldn't take too many people refusing to book with irresponsible companies before changes might be implemented.

(See also p.40 for impact of skiing examples.)

Practicalities

The practicalities of an adventure/activity holiday should not be ignored. The first consideration is whether your guides/instructors are fully qualified in their field. That means not only finding out whether they have been trained, for instance, in climbing instruction, ballooning, or river rafting, but whether they are also fully knowledgeable about the country where the holiday is to take place. A guide who is as new to a country as you are is not likely to be able to point out the pitfalls of religious customs, culture, geographical terrain and physical strain. Look out for companies whose field staff actually live in the countries concerned and have managed to become integrated into the local way of life. These are often the operators who really take their job seriously.

The second consideration concerns the physical demands that are going to be made of you as a tourist and whether operators point these out before you book your holiday, or whether they try to gloss over them for fear of losing your custom. Remember, if one company's tours seem too demanding, there's usually another company more suited to your needs. After all, holidays are still about enjoying yourself and if you choose a company which sets the pace and demands at your level, then the outcome will be a lot of good fun.

Tips

There are a few other factors which any serious adventure/activity tourist should be aware of, regarding both personal safety and the contribution they can make to maintaining a satisfactory level of constructive environmental tourism.

Health should always be a primary concern. It doesn't matter how good your intentions are, they won't prevent disease if you haven't taken the appropriate precautions. Vaccinations and pills such as malaria tablets should be treated seriously. A tetanus jab/booster is also a good idea if you're going to be in an infected area, or even if you're likely to suffer cuts and bruises.

Given the correct advice about how fit you need to be for a holiday, it is not necessarily enough to decide for yourself whether you will cope or not. A medical check-up never goes amiss. Find out what sort of accommodation and living conditions your holiday entails. These should be outlined in company brochures or hand-outs. Ask for more detailed information if you're not satisfied. Try and find out how much good use can be made of local facilities and compare it with what a company is offering. A good starting point for information can be the National Tourist Office of the country concerned. Most of these are situated in London. For complete list contact A.N.T.O.R., 42d Compayne Gardens, London NW6 3RY.

Look into how you can continue contributing to areas of interest after your holiday. There is a wide range of charities and organisations working in diverse fields throughout the world. Ask your tour operator which ones are the most effective and whether the company makes any contribution. This does not always have to take the form of cash donations. For example, Kaleidoscope of India, a small company specialising in Indian tours, contributes by collecting used Ladybird books which they give to Tibetan Refugee Schools; admittedly a very small gesture, but it's more than most are doing and the thought is nevertheless there.

Adventure/Activity Holidays Available

The companies listed here are a selection of those offering the most suitable holidays, combined with a constructive and caring approach to the environment in which those holidays take place. Given the huge market in adventure/activity holidays, they have been split into a relatively large number of sub-categories in order to provide a more ordered picture.

'Adventurous Study'

DISCOVER LTD.
Discover offer a wide range of tours to Morocco, catering for GCSE school trips (group sizes ten to seventy) students, adventure breaks, birdwatching, mule trekking, desert and mountain camping and a host of other alternatives. They aim to give their clients, 'a sound understanding of the different cultures they are visiting, so that their behaviour can be in keeping with it and not offensive to it.' Discover puts approximately 50 per cent of their tour costs back into the local economy. Prices range from £500 to £700 per person for ten days, including return airfare.

Discover also owns The Eagle's Nest Field Study and Activity

Centre in the Cévennes offering accommodation for groups of up to seventy students plus staff. It is situated in a French National Park on the south-facing slopes of Mont Lozère, the highest peak in the Cévennes which form the southern edge of the Massif Central. The Cévennes is an area of outstanding natural beauty, rich in flora and fauna benefiting from Mediterranean sunshine. A short drive away are the breathtaking, but canoeable gorges of the River Tarn, suitable for day trips, expeditions or journeys camping overnight. The area boasts some of the finest caving in Europe and pony trekking or walking through thousands of hectares of spectacular open countryside. Discover aim to give groups/visitors an insight into the diverse human and physical environment the Cévennes has to offer. Prices range from £200 to £300 including coach travel.

Further information from: Timbers, Oxted Road, Godstone, Surrey, RH9 8AD. Tel. 0883 744392.

Ballooning

ADVENTURE BALLOONS.
Just for the experience, or for serious learning, Adventure Balloons offer holidays and courses with their well-trained instructors. Although based in London, flights can be organised throughout the UK and the company's 'Club' run trips to Ireland, France and other European countries. Horse-riding, dry-slope skiing and mountain biking are also offered in conjunction with ballooning trips. Prices range from £100 to £230 for either a single flight or a ballooning weekend. French ballooning holidays start at about £450. Adventure Balloons specify that the minimum age for clients is twenty-one.

Further information from: 3 Queen's Terrace, Hanwell, London, W7 3TS. Tel. 081 840 0108.

Canoeing and Rafting

CANADIAN WILDERNESS TRIPS.
Canadian Wilderness offer clients the opportunity to really get to grips with nature, to experience the 'rough and tumble'

of it. Established in 1972, the company is now a member of the Northern Ontario Tourist Outfitters and the National Association of Canoe Liverers and Outfitters and offers clients the opportunity to experience flatwater and whitewater canoeing in Algonquin Park and Northern Ontario. Prices range from $195 to $700 (about £100 to £350), not including air fares. Accommodation is in bunkhouses/tents and qualified guides are always on hand. Minimum unaccompanied age is eighteen years.

Further information from: 187 College Street, Toronto, Ontario, Canada M5T 1P7. Tel. 0101 416 977 3703.

ENCOUNTER OVERLAND.

As their name suggests, Encounter Overland offer a lot more than just canoeing. They are, however, one of the leading companies working in this field, particularly in Asia. The company has been in operation since 1963 and each year they organise holidays for over 3,000 people. Whitewater river-running in Nepal is offered from October to May, with accommodation in tents beside the river! No experience is required since all trips are accompanied by a fully qualified helmsman, although there is a condition that all clients must be able to swim. Prices range from about £300 to £340 for ten days, not including air fares.

Further information from: 267 Old Brompton Road, London, SW5 9JA. Tel. 071 370 6951 Fax. 071 244 9737 (see also p. 212).

Exploration

DISCOVER THE WORLD.

Specialist operator offering 'comfortable adventures in the natural world' in some of the most remote and beautiful corners of the earth. Impressive range of worldwide destinations, itineraries suited to all ages; expert guides. Sample price: eighteen days in Falkland Islands £3,500.

Further information from: 29 Nork Way, Banstead, Surrey, SM7 1PB. Tel. 0737 373789 Fax. 0737 362341.

EXPLORE WORLDWIDE.

Although Explore offer their clients 'Wildlife and Natural History Adventures', the nature of their trips is better categorised as 'exploration' rather than as 'safari' or 'natural history'. Game viewing in Kenya, gorilla tracking in Zaire and exploring the Okavango Delta are just three of the 'adventures' on offer. Other destinations include Malawi, Tanzania, Madagascar, India, Sumatra, New Guinea and Patagonia, to name just a few. Explore work in small groups, led by professional guides and leaders and are highly respected within the adventure holidays world. Their holidays divide into eight categories: Cultural and Adventure; Wildlife and Natural History; Ethnic Encounters; Easy-Moderate Hiking; Major Treks; Wilderness Experience; Sailtreks and Seatreks; and Raft & River Journeys. Average to lower than average prices. In all, they offer over ninety tours, treks, safaris and expeditions in fifty countries throughout the world.

Further information from: 1 Frederick Street, Aldershot, Hants, GU11 1LQ. Tel. 0252 319448.

GALAPAGOS ADVENTURE TOURS.

A small, specialist operator, the owner of Galapagos Adventure Tours formed the company in 1985, following several years spent living and working all over South America. He has chosen to run tours to his favourite country, Ecuador, taking groups on tailor-made trips. These are escorted by, 'expert graduate leaders, who have lived and worked as naturalist guides in the past.' The emphasis is very much on protection of the environment and the need to minimise the impact of tourism, taking care not to 'interfere with indigenous peoples' lives, avoiding those tours which exploit Indians and supporting those which contribute to a mutual understanding.' Approximately 60 per cent of the income generated is fed back into the economy of the host country. Boats used are small sailing/motor yachts, chosen for their reliability and their ability to sail at night, reaching the outermost islands of the Galapagos. Prices range from £2,500 to £3,500, for about twenty days.

Further information from: 29 Palace View, Bromley, Kent, BR1 3EJ. Tel. 071 403 8663 Fax. 081 460 7908.

Also running to the Galapagos is Alfred Gregory Photo & Trekking Holidays: Woodcock Travel, 25/31 Wicker, Sheffield, S3 8HW. Tel. 0742 729428. Alfred Gregory offers special de luxe cruises around the Galapagos by private yacht. Their clients go alone into villages to seek photographic locations and observe local traditions at all times.

PASSAGE TO SOUTH AMERICA LTD.

Exploration of the cities, towns, mountains and lost villages of South America is the aim of this company, who seek to develop low-impact tourism. Also available is the opportunity for skiing in the Andes. All tours are to underdeveloped and developing countries and ground operations are handled by 100 per cent locally owned companies who receive, on average, 60 per cent of the tour cost. The other 40 per cent is paid in London to the airline, which is often South American. Itineraries are tailored to the individual in an effort to allow a real understanding of the host country. Prices range from £1,100 to £2,700 for fourteen to twenty-one days.

Further information from: 41 North End Road, West Kensington, London, W14 8SZ. Tel. 071 602 9889 Fax. 071 371 1463.

REEF AND RAINFOREST TOURS LTD.

It is difficult to categorise Reef and Rainforest Tours due to the diversity of its trips. 'Exploration' is certainly applicable, but so too are walking, wildlife and natural history. Tours to Belize and other parts of South America are organised, some accompanied by the director, who is an expert in this part of the world. Itineraries cover everything from 'the Belize Adventurer', seventeen days exploring the Mayan ruins, caves, rainforest and sailing the Reef, to the nineteen- day 'Christmas and Cocos (Keeling) Islands' natural history tour. Other trips include 'the falls, whales, glaciers and natural history' of Argentina and exploring the habitats of Costa Rica. Prices range from £1,600 to £3,000 for eleven to nineteen days. Details of their policies are included in Chapter 9 under 'Wildlife/Ecology/Natural History'.

Further information from: 205 North End Road, London, W14 9AP. Tel. 071 381 2204 Fax. 071 386 8924.

WORLDSAWAY
Adventure tours in India, the Far East and South America, including Himalayan trekking. Pre-departure dossiers available. Average price: The Annapurna Explorer – sixteen days for £1,650.

Further information from: 101 Eden Vale Road, Westbury, Wiltshire, BA13 3QX. Tel. 0373 858956 Fax. 0373 858351.

Mountain Travel

EXPLORASIA LTD.
Much of ExplorAsia's work is based on trekking in Nepal, but their expeditions cover too wide a field to merit such a narrow categorisation. These are treks with a definite theme or emphasis, covering aspects such as photography trips, river-running, game-viewing and – the title says it all – a 'Rhododendron Trek – Springtime in Nepal'. Trips are also organised to Ecuador and the Galapagos Islands.

A good reflection of ExplorAsia's attitude towards conservation can be seen in the fact that they employ approximately 1,000 local staff, the majority of whom are retained during the quiet summer months. The company is also involved with a number of conservation projects and include in their prices a fee which is contributed to the King Mahendra Trust for Conservation, which covers the whole Annapurna area. Future legislation might involve a reduction in current charges, to be replaced by a standard Conservation Project charge of US$30 per trekker, regardless of trek length. Prices range from £1,700 to £3,000 for about twenty days.

Further information from: 13 Chapter Street, London, SW1P 4NY. Tel. 071 630 7102 Fax. 071 630 0355.

Multi-Activity

AMERICAN ADVENTURES.
Camping, walking, diving, cycling, horse-riding, river rafting or holidays on a working ranch are among the options offered by this specialist in Canada, the US, Alaska and Mexico. Groups are small and prices very competitive. Twenty-eight days 'American Explorer' tour £499 less flights.

Further information from: 45 High Street, Tunbridge Wells, Kent TN1 1XL. Tel 0892 511894; Fax. 0892 511896.

ASSOCIATED ACTIVITY HOLIDAYS.

Archery, sailing, windsurfing, cycling, riding, flying, sub-aqua for adults only. From £235 per week.

Further information from: PO Box 398, Poole, Dorset BH15 4JP. Tel. 0202 4181.

CALSHOT ACTIVITIES CENTRE.

One of Britain's largest outdoor centres offering residential or non-residential courses in many water and land-based activities. Situated just west of the New Forest. Offers an Environmental Studies programme with day and weekend courses in subjects such as 'The Flora and Fauna of the New Forest' and 'Life Along the Shoreline'. Average price weekend £85 residential (all meals). School courses run through the week.

Further information from Calshot Spit, Fawley, Southampton, SO4 1BR. Tel. 0703 892077.

COUNTRYWIDE HOLIDAYS ASSOCIATION.

Country-house accommodation as a base to activity holidays at thirteen UK centres. Everything from dancing to countryside appreciation and outdoor activity weekends.

Further information from: Birch Heys, Cromwell Range, Manchester, M14 6HU. Tel. 061 225 1000 (see also p. 228).

HEADWATER HOLIDAYS.

Headwater operate only in France, where canoeing, cycling, riding, skiing, windsurfing and dinghy sailing are all on offer under the eye of qualified instructors. They do not take guided groups, which they regard as intrusive, but allow clients to, 'make their own way, in 2's and 4's, or with their friends'. Headwater make an effort not to 'swamp' the places they visit by seldom having more than three or four rooms booked at any one hotel and will only accept bookings on the condition that local codes of practice, as outlined by a representative on your arrival, are adhered to. Their tours are highly commended, combining low-impact facilities with interesting locations. Prices range from £250 to £700 for six to fourteen nights, depending on whether you travel by rail or drive yourself.

Further information from: 146 London Road, Northwich, Cheshire, CW9 5HH. Tel. 0606 48699 Fax. 0606 48761.

THE NATIONAL CENTRE FOR MOUNTAIN ACTIVITIES.

A large centre offering courses from two days to ten weeks. Mountaineering, outdoor photography, climbing, skiing (dry slope and snow), orienteering, canoeing. All groups and ages catered for.

Further information from: Chapel Curig, Gwynedd, North Wales. Tel. 06904 214/280.

OUTFACE ADVENTURES.

Outface Adventures, run from a seventeenth-century farmhouse hotel on the edge of the Wye Valley, offer active adventurers the opportunity to indulge in a wide variety of outdoor pursuits, from walking to abseiling, in the Welsh countryside. Expert tuition is on hand at all times, and clients are treated to the sort of active weekend that they never thought they could manage. Prices vary, but a weekend break costs about £250 (all inclusive) while a week's course costs about £390.

Further information from: PO Box 14, Monmouth, Gwent, Wales, NP5 4YW. Tel. 0600 83482.

OUTWARD BOUND TRUST.

Outward Bound courses are the epitome of adventure/activity holidays. They are also one of the original organisations to offer such holidays and, like so many good ideas, owe their success partly to the fact that their format has remained true to the 'simple' style of the early days. There is nothing flash or luxurious about Outward Bound; it aims to provide healthy, adventurous, stimulating and challenging courses and does so with the minimum fuss. Virtually every type of outdoor activity imaginable is covered, available from the south of England to the north of Scotland and at centres throughout the world. Education and instruction on living outdoors is perhaps the best way to sum up what the Outward Bound offer. They are established pioneers in the art of conserving the environment while putting it to the best use possible. Prices vary throughout the year but two typical examples are an eight-day course in Wales in February for £305 and a twenty-day course in Scotland in July/August for £635.

Further information from: Chestnut Field, Regent Place, Rugby, CV21 2PJ. Tel. 0788 560423 Fax. 0788 541069.

SCOTTISH YOUTH HOSTELS ASSOCIATION.
Holidays in the Scottish countryside offering a wide range of outdoor activities – cycling, pony trekking, golfing, water sports, sports and adventure.

For further information: 7 Glebe Crescent, Stirling, FK8 2JA. Tel. 0786 51181.

WORLD CHALLENGE EXPEDITIONS.
After attending a selection weekend, a place on an Expedition Team may be allocated. Expeditions, carried out for genuine research and study reasons, take place worldwide, from Borneo to Canada. Fees are in the order of £1,800 for a month. Suitable for those completing their Duke of Edinburgh's Gold Award.

Further information from: Walham House, Walham Grove, London, SW6 1QP. Tel. 071 386 9828.

YOUTH HOSTELS ASSOCIATION.
A wide range of activity and adventure holidays based in the YHA hostels across England and Wales. Walking, riding, watersports, sand yachting, air sports, rock sports, special interest breaks, etc.

Further information from: Trevelyan House, 8 St Stephen's Hill, St Albans, Herts. AL1 2DY. Tel. 0727 55215.

Overland Holidays

ARCTIC EXPERIENCE.
Arctic Experience are specialists in multi-activity holidays in Iceland, Greenland, Norway, Sweden and Canada. Walking, trekking, river-rafting, horse-riding, canoeing, kayaking and skidoo trips are all available (see also Chapter 9). Arctic Experience work in the belief that, 'tourism can benefit conservation by promoting a better understanding of the many threats to the natural world'.

Activity/Adventure Holidays 213

They also accept that 'it is important for tourism to be socially responsible and environmentally conscious' and are trying to reduce their group sizes in an effort to reduce the impact on the environment. Founder members of Green Flag International (see p.226). Their range of tours is extremely comprehensive and prices vary considerably. The fourteen-night 'Walking on the Wild Side' trip to Iceland costs from £930 to £950, while a seventeen-night holiday in Alaska and Yukon costs about £2,820.

Further information from: 29 Nork Way, Banstead, Surrey, SM7 1PB. Tel. 0737 362321.

DRAGOMAN.

A dragoman was a local guide and agent hired by early travellers to Asia, Africa and the Middle East. Such is the role of the company today who offer overland expeditions across Asia, Africa, the Middle East and South America. These expeditions range from two weeks (for those with limited time), right through to thirty-one weeks, but many of their trips are five weeks. Travel is in specially designed Mercedes expedition vehicles and accommodation is mainly camping with occasional stays in local hotels. In South America hotels are used more frequently. Groups prepare their own food and the cost usually works out between £120 and £200 per week (depending on length of trip), fully inclusive of land content. Three and a half weeks in India is roughly £500 while twenty-four weeks right around South America is roughly £2,600.

Further information from: Camp Green, Debenham, Suffolk IP14 6LA. Tel. 0728 861133 Fax. 0728 861127.

ENCOUNTER OVERLAND.

In its extensive programme, Encounter Overland features adventure travel worldwide, ranging from Brief Encounters that last ten days to a transcontinental expedition that takes twenty-nine weeks. The aim of the Brief Encounters is to make adventure travel accessible to those people with restricted holiday time. Included in the brochure are Brief Encounters to destinations such as Peru, Nepal, Turkey, East Africa, Egypt and Tibet. Prices for these trips start at £390, excluding flights. The cost includes transport, accommodation and all meals.

Prices for the transcontinental expeditions lead in at £975. This is the cost of the overland journey from Tangiers to Dakar, taking six weeks and three days and crossing Morocco, Algeria, Mali and finally ending up in Senegal. The price includes all accommodation, food and transport excluding flights.

Encounter Overland is market leader in overland expeditions and the company had been in operation for over twenty-seven years.

Further information from: Encounter Overland, 267 Old Brompton Road, London, SW5 9JA. Tel. 071 370 6845 Fax. 071 244 9737.

EXODUS EXPEDITIONS.

Exodus Expeditions offer a vast network of holidays worldwide: safaris, hiking, sailing, trekking, all things energetic and all in true adventure spirit. Although Exodus cannot say what percentage of their generated income is fed back into the local economy, they do have a 'policy of using local hotels, guides, services, etc. where possible, rather than being either totally self-contained or using multinational facilities'; however, they are concerned to run tours of a 'non-invasive nature' which deal with the wildlife, culture and history of places visited. They are highly rated by trekkers worldwide. Prices range from £600 (for Poland) to £2,000 (for Galapagos).

Further information from: 9 Weir Road, London, SW12 0LT. Tel. 081 675 5550 Fax. 081 673 0779.

GUERBA EXPEDITIONS.

Established in 1977, Guerba now offers adventure holidays, overland expeditions and safaris throughout Africa, arranging over 3,000 holidays each year to a total of twenty-seven African countries. 'Holidays' last from one to four weeks, while 'expeditions' last from seven to nineteen weeks. Guerba offer one of the most comprehensive range of African holidays on the market; they have forty itineraries and can be considered specialists in their field. Good pre-departure briefing notes. Prices range from £300 for two weeks in Morocco to £2,800 for the mammoth twenty-seven-week 'Africa all the way' expedition.

Further information from: 101 Eden Vale Road, Westbury, Wiltshire, BA13 3QX. Tel. 0373 826611 Fax. 0373 858351.

JOURNEY LATIN AMERICA.
Journey Latin America have been organising specialist tours to this part of the world since 1980. Small escorted tours, tailor-made itineraries and budget-travel flights are all available. Destinations include Argentina, Antarctica, Brazil, the Falklands, Guatemala, Mexico, Panama, Santo Domingo and Venezuela, plus many more. They offer environmental expeditions into South America and have dedicated a brochure to 'Small Groups'. Seven nights in Peru on the trail of the Incas costs about £650, not including flights.

Further information from: 16 Devonshire Road, Chiswick, London, W4 2HD. Tel. 081 747 3108 Fax. 081 742 1312.

ROAMA TRAVEL.
Specialising in tours to Nepal, India and Kenya where treks, river rafting or climbing can be enjoyed. Use facilities provided by small hotels and teahouses on the main trails. Groups or individuals catered for. Average price – Classic Everest, twenty-eight days £1,625.

Further information from Shroton, Blandford Forum, Dorset DT11 8QW. Tel. 0258 860298.

SOUTH AMERICAN EXPERIENCE LTD.
South American Experience offer both overland and trekking holidays. They state they are 'concerned about the preservation of the South American rainforest, its fauna, flora and its tribal peoples' and have 'joined together with Survival International in a project to help save and conserve the Amazon.' The company offers reduced fares and credit facilities to representatives of aid agencies, as well as playing a significant role in projects such as Programme for Belize, founded to preserve and endow over 250,000 acres of rainforest. Mexico, Peru, Brazil, Ecuador and Central America are all featured in their brochure. Prices range from £450 to £800 for fifteen to twenty-one days, not including air fares.

Further information from: Garden Studios, 11–15 Betterton Street, Covent Garden, London, WC2H 9BP. Tel. 071 379 0344 Fax. 071 379 0801.

TEMPLE WORLD/ROYAL GEOGRAPHICAL SOCIETY TOURS.
Tours led by eminent scholars and explorers. Pre-departure lectures. Various tours arranged. African tours for '91–'92 include Zimbabwe, with Victoria Falls and Okavango Delta for £2,275, and The Rift Valley and Masai Mara, Kenya for £1,920. Further information from: Temple World, 3–4 St Andrew's Hill, London, EC4V 5BY. Tel. 081 940 4114.

TOP DECK.
The current management of this company all started 'on the road, as couriers and drivers . . . in 1972'. Top Deck offers trips from London, travelling throughout South East Asia and South America, specialising in every country visited. '100 per cent of on-the-road-tour funds are spent locally', while '100 per cent of on-the-road-trip costs are fed into the local economy'. For every worldwide booking, £1 is donated to help educate and aid Nepalese children, while an annual Top Deck Ball and other fund-raising events are also organised throughout the year. Top Deck's trips are as diverse and varied as can be imagined, covering areas such as London to Kathmandu in eleven weeks, African safaris and Zambezi white-water rafting. Prices range from £400 to £2,500 for fourteen to seventy days, not including flights. Top Deck specify that their tours are for eighteen to thirty-five-year-olds.

Further information from: Top Deck House, 131/135 Earl's Court Road, London, SW5 9RH. Tel. 071 244 8641 Fax. 071 373 6201.

TRACKS.
Tracks offer a wide variety of holidays throughout Europe and Africa. In Europe they mainly organise camping trips; in Africa they offer adventure safaris, 'to suit all tastes and budgets', from one to twenty weeks. Camping safaris, lodge safaris, gorilla and chimpanzee searching are just a few of the

highlights. Fifteen weeks across Africa by truck costs about £2,000.

Further information from: 12 Abingdon Road, London, W8 6AF. Tel. 071 937 3028.

Sailing/Flotilla Sailing

FALCON SAILING.
Included in the price of Falcon Sailing's holidays is instruction from qualified members of the Royal Yachting Association. Dinghy sailing, windsurfing, sailing cruiser instruction and flotilla holidays are all available in Greece, Turkey and Sardinia. Prices range from £300 to £600 for fourteen days. No experience required.

Further information from: 13 Hillgate Street, London, W8 7SP. Tel. 071 727 0232 Fax. 071 727 3142.

MALTA CRUISING SCHOOL LTD. (Formerly FOWEY CRUISING SCHOOL.)
The only school on the island approved by the Royal Yachting Association and the only one licensed by the Maltese government. They are also members of the National Federation of Sailing Schools. As well as Malta, Tunisia and Sicily are featured in their brochure. Price is set around £500 and no experience is required.

Further information from: 4, Forth Flat, 187 Marina Road, Pieta, Malta. UK office – 32, Fore Street, Fowey, Cornwall PL23 1AQ. Tel. and Fax. 0726 832129.

OTHER COMPANIES.
Clearwater Holidays, Tel. 0926 450002; Flotilla Sailing Holidays, Tel. 081 969 5423; Headwater Holidays, Tel. 0606 48699; Sunsail International, Tel. 0705 210345; Turkish Delight Holidays, Tel. 081 891 5901; Western Isles Sailing and Exploration (see p.159).

Skiing

ABERNETHY OUTDOOR CENTRE.
At Abernethy, an independent Christian Trust, skiing is the winter priority. Skiing courses for all standards are offered, with

qualified tutors available. Cross-country skiing is also organised. Seven nights between January and May costs about £180 plus VAT, all-inclusive. The staff are all committed Christians and are keen to share their faith with their guests.

Further information from: Nethy Bridge, Inverness-shire, Scotland, PH25 3ED. Tel. 047 982 279.

CALSHOT ACTIVITIES CENTRE.

Calshot Activities Centre, recognised by the English Ski Council, offers weekend and evening courses for all standards on the largest indoor ski slope in Britain. Qualified instruction is included in the price of about £70 for a weekend, which also covers accommodation, meals and equipment.

Further information from: Calshot Spit, Fawley, Southampton, SO4 1BR. Tel. 0703 892077/891380.

CANADIAN WILDERNESS.

Cross-country skiing in Algonquin Park is offered by Canadian Wilderness Trips, as well as dog-sledding and snow-shoeing. All meals and accommodation are included in the price, but cross-country skis are not, although they are available for hire. Holidays are organised for both individuals and groups and itineraries can be tailor-made.

Further information from: 187 College Street, Toronto, Ontario, Canada, M5T 1P7. Tel. 0101 416 977 3703. Fax. 0101 416 977 7112.

NORWEGIAN MOUNTAIN TOURING ASSOCIATION.

The Norwegian Mountain Touring Association has been in operation since 1868 and today organises holidays for 5,000 people annually. The association specialises in cross-country skiing in the mountains of Norway for both groups (maximum twenty) and individuals. Accommodation is in dormitories and all meals are provided.

Further information from: Den Norske Turistforening, Boks 1963, Vika N-0125, Oslo 1, Norway. Tel. 010 47 2832550.

Other companies include Headwater Holidays (p.209); and Sherpa

Expeditions, 131a Heston Road, Hounslow, TW5 0RD. Tel. 081 577 2717.

Walking & Trekking

BORDERLINE.

Mountain walking holidays in the Ordesa National Park in the Pyrenees, run by a locally-based Englishman living in Barèges. Accommodation in mountain refuges, fauna and flora studies from local botanists and ornithologists, and inspiring itineraries. Mountain biking around Barèges also possible and winter sports packages also available. Average two-week trip £330, not including transport.

Further information from: Peter Derbyshire, Les Sorbiers, rue Ramond, 65120 Barèges, France. Tel. 01033 6292 6895 Fax. 01033 6292 6693.

CLASSIC NEPAL.

Specialist trekking company to Nepal, Himalayas. Direct sell. Average price: Annapurna Sanctuary Tour £715.

Further information from: 33 Metro Avenue, Newton, Derbyshire DE55 5UF. Tel. 0773 873497 Fax. 0773 590243.

CRYSTAL HOLIDAYS.

Hillwalking in Austria from mountain hut to hut is an option from Crystal's 'Mountain Action' programme. Suitable for those not necessarily super-fit, it offers low-impact holidays in the Kaprun area taking in the beauties of the Kitzsteinhorn to the Grossglockner. Groups of eight maximum; regional departures. From £449 a week.

Further information from: Crystal House, The Courtyard, Arlington Road, Surbiton, Surrey, KT6 6BW. Tel. 081 390 3335/081 399 5144.

DICK PHILLIPS.

'Go Icelandick' with this excellent specialist who, 'Deals only with Iceland, but makes no attempt to market the country.' The policy, formed in 1962, is, 'To assist walkers, cyclists, mountaineers and kindred travellers . . . to get the most from their trips to Iceland, without adding to their expenses. The principles of self-help will come naturally to those who find Iceland most attractive and it is not his (Dick Phillips') policy to provide services of a type which would tend to change the essential character of this country.' Dick Phillips stock the largest range of maps and other publications on Iceland available in Britain. Prices range from £310 to £886 for twelve to fifteen days.

Further information from: Whitehall House, Nenthead, Alston, Cumbria, CA9 3PS. Tel. 0434 381440.

EXODUS EXPEDITIONS.

An excellent walking, adventure and overland specialist for small groups with sustainable policies, fusing local facilities and guides throughout. Good pre-departure briefing notes. Re-cycled brochure!

Particular interest holidays also offered – wildlife, botany, anthropology, etc. Has walking programme and adventures programme covering forty countries.

Further information from: 9 Weir Road, Balham, London, SW12 0LT. Tel. 081 673 0859 Fax. 081 673 0779.

HF HOLIDAYS.

UK market leader in walking holidays, running for over a hundred years, HF offers an extensive range of holidays. They own nineteen country houses in coastal towns and national parks which act as UK bases. European and worldwide tours also available, and low-impact facilities stressed throughout. Prices from £199 for seven nights full board.

Further information from: Imperial House, Edgware Road, London, NW9 5AL. Tel. 081 905 9556.

HIGH PLACES LTD.

High Places has only been running trips since 1986, but already they are well established as responsible operators, 'organising treks to remote and unspoilt parts of the world'. India, Sinai, Ecuador, Iceland, Tasmania, Borneo, North Korea and Egypt are all on offer in their brochure which also includes excellent hints on which trip suits which age-group and the different categories of treks available: 'Expeditions, tough treks, moderate treks, and short treks and walks.' High Places say that, 'People, customs and local food are the essence of foreign travel . . . We try hard to be sensitive to the culture and the environment of the countries we visit.' The company also runs a slide presentation and travel shop around the country. Prices range from £800 to £2,200 for nine to twenty-one days.

Further information from: The Globe Works, Penistone Road, Sheffield, S6 3AE. Tel. 0742 822333 Fax. 0742 820016.

KALEIDOSCOPE OF INDIA LTD.

Kaleidoscope of India is a small company offering personalised and specialised tours to India. These fall into six categories: walking; culture and religion; wildlife; festivals; healing; and architecture. Most tours are accompanied by the director, whose concern about India's conservation and wildlife is of central importance. She says, 'Specialising in India gives me great scope to educate clients on the preservation of the environment and its people.' Visits to wildlife parks are included whenever possible, as are talks on threatened species. The director is a Fellow of the Royal Geographic Society and a member of both the Fauna and Flora Preservation Society and the International Snow Leopard Trust. Groups range from eight to twenty people. Prices range from £1,700 to £2,200 for fifteen to twenty days.

Further information from: 70 Chalk Farm Road, London, NW1 8AN. Tel. 071 485 1444 Fax. 071 485 1470.

REMOTE TRAVEL COMPANY.

Walking holidays in Greenland, meeting Eskimos and enjoying the Arctic summer are among trips on offer. Small groups led by Arctic explorer.

Further information from: High Leaze, Oare, Marlborough, Wiltshire. Tel. 0672 63997.

SHERPA EXPEDITIONS.

In business since 1973, Sherpa offer an impressively comprehensive range of destinations throughout the world. Nepal, Tibet, Ladakh, China, South America, Africa, Morocco, Spain, Greece,

Turkey and the Alps are all included in their brochure. Sherpa are keen to make walks as comfortable as possible, using porters and pack animals to carry bags, leaving you free to enjoy the experience. Their Hotel Treks programme is perfect for the older traveller. Twenty-one days spent around Everest costs about £1,315. They also now organise cross-country skiing in France in winter.

Further information from: 131a Heston Road, Hounslow, Middlesex, TW5 0RD. Tel. 081 577 2717 Fax. 081 572 9788.

THE SURVIVAL CLUB.
They cleaned up Base Camp at Everest – the world's highest rubbish dump – and they continue to make conservation issues high on their list of priorities. They lead expeditions to Borneo, Europe, Pakistan and others, and they trek, canoe, pothole and ski. Annual membership £25, thereafter newsletters, news of trips, lectures etc. await.

Further information from: PO Box 156, Southport, PR9 9GS. Tel. 0704 211708.

Other walking/trekking companies include Brochure Roama Travel: Larks Rise, Shroton, Dorset, DT11 8QW. Tel. 0258 860298, for Nepal, India and Kenya. Noted for their environmental awareness are Himalayan Travel: Nurse's Cottage, Long Lane, Peterchurch, Hereford, HR2 0TE. Tel. 0981 550246, for Nepal. Himalayan Kingdoms: 20 The Mall, Clifton, Bristol BS8 4DR. Tel. 0272 237163; Fax. 0272 744993.

Wilderness

ACCESSIBLE ISOLATION HOLIDAYS.
This is a company who have taken a good look at the country in which they offer holidays – Canada – and have assessed how best to make the most of what's available. At the same time as instructing whatever activity is chosen, they educate their clients about the Canadian wilderness, its wildlife and how to treat it while there. Virtually every type of outdoor activity is on offer, all complete with professional leadership, local guides and specialist equipment where necessary. Horse-riding, wilderness lodges, camping, and

float planes are just a few of the attractions available, as well as Canadian safaris, whale-watching and even a 'Polar Bear Expedition'. Prices range from £350 to £6,000 for three to sixteen days.

Further information from: Midhurst Walk, West Street, Midhurst, West Sussex, GU29 9NF. Tel. 0730 812535.

Accessible Isolation also organise holidays at the Temagami Wilderness Centre. Temagami offers much the same type of holiday and in a similar 'educational' spirit as Accessible Isolation, practising 'minimal impact and no trace camping techniques' and utilising 'local native culture as an integral part of the wilderness experience. 'They also support a national canoe route clean-up programme headed by the Canadian Recreational Canoeing Association and cooperate with the Provincial Ministry, of Natural Resources in providing information.

Further information from: Accessible Isolation or 210–360 Billings Court, Burlington, Ontario, Canada, L7N 3N6. Tel. 0101 416 632 8124; Fax. 0101 416 632 4520.

CANADIAN WILDERNESS.

Local agent offering canoe travel in the remote areas of Canada. Based on Kowawaymog Lake in the Algonquin Park, these are low-impact, camping-based holidays. An outdoor school offering woodsmanship, writing, painting and photography classes also exists here.

Further information from: 187 College St, Toronto M5T 1P7 Canada. Tel. 0101 416 977 5763; Fax. 0101 416 977 7112.

Chapter 12

Conventional Holidays

In a book attempting to analyse the flaws of mass tourism you might expect 'conventional' holidays to be discouraged in favour of the type of holidays listed in the other chapters. However, as we've tried to explain in chapters 5 and 6, there is an important place for *mass* tourism which is needed to cope with the huge numbers of people who want to and have the resources to go on holiday at least once a year.

Undoubtedly the nature of mass tourism is changing. In the past people have been prepared to put up with holidays in newly-built resorts with all the attendant problems in relation to quality but recent trends in the holiday market reflect change. The traditional fortnight package to a Spanish or Greek coastal or island resort has become considerably less popular. However package holidays are still extremely important, accounting for about two-thirds of the overseas holiday market, with the balance made up by independent holidaymakers. What is happening is that the 'package' is evolving to meet new demands – there are now a greater range of options open to the holidaymaker and the signs are that the trend is set to continue.

The decision of Thomson's, Britain's biggest tour operator, to radically alter its package holiday approach is a sign of the market's recognition of change.

The package has undoubted attractions. It provides reasonably affordable holidays to a large number of people at a (usually) predictable standard and quality. What is happening is that the demand for something different is spreading the tourist load over a wider number of countries and resorts. While much of this is to be welcomed, it is critical that such dispersal is conducted in such a way as to conserve the attractions of the new destinations, rather than turn them into imitations of the resorts now desecrated.

In the early 1960s and 1970s the belief was that independent holidays would increase as holidaymakers, weaned on package holidays, gained in confidence and branched out on their own. Well, independent holidays have grown, but interestingly not as fast as the package market. The holidaymakers with the greatest disposable income tend to be those people with the least time to organise intricate travel arrangements, and so they pay for the convenience and time-saving of the organised holiday package. The packages available reflect the market and everything from trips on Concorde to the Orient Express now come packaged – a package no longer means just two weeks on a Costa.

Unlike other chapters, here we are not recommending particular companies or holidays available. There are simply too many and it would be impossible to analyse the several thousand options open in any worthwhile detail. Rather we recommend you do your own environmental audit, as outlined in Chapter 6, on any holiday you are considering. A useful back-up to this, however, to ensure you choose the most sensitively designed holiday, is Green Flag International – the scheme based on a kite-marking system which environmentally audits holidays, tour operators and resorts. These are analysed in detail according to sustainable tourism criteria, which include:

- the regard paid to landscape, wildlife and cultural heritage
- energy efficiency
- waste disposal and re-cycling
- interaction with local communities in terms of goods and services
- sympathetic building and architecture

Initial entry into the scheme by tour operators is by subscription on a voluntary basis, in the same way as the Blue Flag Scheme for beaches; the hope is that this operation will grow in size and importance until all operators, large and small, feel it necessary to be members – something they can only achieve by passing the test. Then the 'good tourist' can enjoy a 'good holiday' offered by a 'good tour operator' and the travel industry can look forward to a long and successful future.

Association of Independent Tour Operators (AITO)

The Association was founded in 1976 mainly in response to the problems posed for smaller travel companies by a sudden sharp increase in bonding requirements following a couple of major collapses. The majority of AITO companies are small and owner-managed with consequent advantages over mass-market operators in the area of product knowledge and personal service. AITO currently has around 70 member companies and accounts for between 10 and 15 per cent of the package market.

Recently AITO has forged strong links with Green Flag International. The Chairman of AITO, Noel Josephides, has stated:

> "The more farsighted of AITO's members have begun to realise that it is time to look and plan ahead, to protect the product on which our businesses depend.
>
> We are not idealists or dedicated conservationists. We are businessmen working in one of the most competitive and high-risk of industries, who have realised that, if we do not help conserve the very places which our clients clamour to see, then, in five years' time, we shall have no clients at all – those dream locations will have been ruined.
>
> Through Green Flag, we can seize the opportunity to begin to impose a conservation discipline on ourselves and, in turn, on the host countries receiving our clients. Such an innovative undertaking will not be without its teething problems, but we are determined to support Green Flag's aims actively and to help guide it towards safeguarding the future of our industry."

For further information contact AITO , PO Box 180, Isleworth, Middlesex, TW7 7EA Tel. 081 569 8092 Fax. 081 568 8330.

GREEN FLAG

Tour Operators currently (1992) supporting the Green Flag initiative are:

Allegro Holidays, Arctic Experience, Barefoot Traveller

Ltd, Barn Owl Travel, Beach Villas, Bird Holidays, BTCV Natural Breaks, Countrywide Holidays, Cox & Kings, CV Travel, Eurocamp, Filoxenia, Greco File, Greek Islands Club, Island Holidays, Moswin Tours, NSR Travel, Pure Crete, Saga Holidays, Simply Crete, Sun Blessed Holidays, Sunvil Holidays, Travel Club of Upminster, Turkish Delight Holidays, VFB Holidays, Wildlife Travel and Zoe Holidays.

ALLEGRO HOLIDAYS.
The company offers hotels and villas of 'real character' in the Costa del Sol, Neapolitan Riviera, Sicily, Sardinia, Corsica, the Algarve and Tenerife. They also offer manor houses in Portugal, paradores and 'country hotels' in Spain and historical hotels in Tuscany and Umbria. A third programme features India, Sri Lanka, Egypt and Nile cruises.

Further information from: 15a Church Street, Reigate, RH2 0AA. Tel. 0737 244870 Fax. 0737 223590.

ARCTIC EXPERIENCE.
Arctic Experience specialises in holidays to the Arctic and sub-Arctic regions. The main destinations are Iceland, Greenland, Spitsbergen, Norway, Sweden, Finland, Denmark and Canada. Ninety per cent, approximately, of the holidays they arrange are tailor-made. They do not accept bookings from travel agents. (See p. 211 for more details.)

Further information from: 29 Nor Way, Banstead, SM7 1PB. Tel. 0737 362321 Fax. 0737 362341.

BAREFOOT TRAVELLER LTD.
Run by a Trinidadian couple, this company specialises in birdwatching in Trinidad and Tobago. A fourteen-day package would cost approx £1,400, including seven nights in the ASA Wright Nature Centre, with all tours and full board, and staying in cottages in Tobago with a cook and tours round the island included.

Further information from: 13 Millpond Court, Bourneside Road, Addlestone, Surrey, KT15 2JA. Tel. 0932 845589 Fax. 0923 845772.

BARN OWL TRAVEL.
Barn Owl Travel have been in operation since 1973 and are specialists in the birds of the British Isles. Other destinations include Austria, France, Iceland, Israel, Menorca, Spanish Pyrenees and the Netherlands. Holidays are arranged for individuals and for groups of up to eight and many are led by the director whose main interests are birds and butterflies.

Further information from: 21 Heron Close, Lower Halstow, Sittingbourne, Kent, ME9 7EF. Tel. 0795 844464. (See also p.153).

BEACH VILLAS.
An independent family-owned company, offering villa and apartment holidays throughout the Mediterranean and other European holiday places.

Further information from: 8 Market Passage, Cambridge, CB2 3QR. Tel. 0223 311113 Fax. 0223 313557.

BIRD HOLIDAYS.
This independent company was set up to raise funds for the Royal Society for the Protection of Birds. Expert leaders, mainly RSPB staff, lead tours to Corfu, Crete, Hong Kong, Iceland, Majorca, Spain, Tunisia and Turkey as well as to RSPB reserves in Britain. Accommodation is normally in 3- or 4-star hotels with daily excursions by coach or on foot. Prices range from £500 for one week in Tunisia to £1,200 for two weeks in Iceland. Staff and leaders are involved in action to protect sites and influence conservation attitudes in the countries they visit.

Further information from: Dudwick House, Buxton, Norwich, NR10 5HX. Tel. 0603 278296.

BRITISH TRUST FOR CONSERVATION VOLUNTEERS NATURAL BREAKS.
Almost 600 holidays on offer around the British Isles. Also International Working Holidays (see p.120).

Further information from: Room GT, 36 St Mary's St, Wallington, Oxon, OX10 0EU. Tel. 0491 39766.

COUNTRYWIDE HOLIDAYS.

Countrywide Holidays (CHA) has been offering 'green' holidays for a hundred years. The company was started by Revd T.A. Leonard in 1891, as an alternative to the usual seaside holidays common at the time. He first took a party of thirty people to the Lake District, to educate them about the countryside and wildlife. His walking parties became so popular that the company has grown, and today owns thirteen country houses around Britain, from which guided walks are taken. The paths used are rotated from week to week in order to minimise erosion. Green Flag International is currently carrying out an environmental audit, to assess and advise on how CHA can make its houses as environmentally friendly as possible. An investment programme is underway to upgrade many of the houses, and any environmental changes which need to be made will be incorporated into this. The food on offer is good and wholesome, with vegetarians and those on special diets catered for.

The houses are located in the Lake District, Peak District, West Country, Isle of Wight, Scottish Highlands and North Wales. In addition to the normal walking weeks, special interest holidays are available, including landscape painting, photography, birdwatching and countryside appreciation. The average cost for a week is £175–£205 full board.

Further information from: Dept GT, Birch Heys, Cromwell Range, Manchester, M14 6HU. Tel. 061 225 1000.

COX AND KINGS TRAVEL.

Set up in 1758, this company offers special interest tours to several long-haul destinations. Regarded as one of the UK's most professional and environmentally aware operators. (See p. 164 for detailed entry.)

Further information from: St James Court, 45 Buckingham Gate, London, SW1E 6AF. Tel. 071 931 9106.

CV TRAVEL.

Originally set up in 1972 as Corfu Villas, the company remains small but with a good reputation for standards, personal attention and service. It operates villas in Tuscany, southern Italy, Majorca, the south of France, Greek Islands and the Algarve. In winter it runs to several destinations including the Caribbean, the Far East, the Seychelles and Mauritius, 'concentrating on individual, owner-managed hotels'.

Further information from: 43 Cadogan Street, London, SW3 2PR. Tel. 071 581 0851 Fax. 071 584 5229.

EUROCAMP TRAVEL.

The company has been offering self-drive holidays to hundreds of ready-erected tents and mobile homes throughout Europe since 1973. Highly respected and a good quality product is offered.

Further information from: 28 Princes Street, Knutsford, WA16 7BR. Tel. 0565 633844.

FILOXENIA LTD.

This is the tailor-made holiday division of Greco File (see below). It concentrates on providing specific holidays to suit a traveller's requirements, mostly in mainland and island Greece. Holidays provided take account of local environmental issues and the company makes use of existing accommodation within the original infrastructures.

GREEK ISLANDS CLUB.

Greek island villa and sailing holiday specialist. Very highly rated and committed to sustainable tourism. A varied and impressive programme. Specialises in Ionian Isles. Works closely with Friends of the Ionian and actively tries to preserve the culture and environment of these holiday islands.

Further information from: 66 High Street, Walton-on-Thames, KT12 2BU. Tel. 0932 220477 Fax. 0932 229346.

GRECO FILE.

Consultancy to Greece offering advice on products available in smaller operators' brochures. Promotes only smaller operators and offers free advice line. Environmental considerations taken into account; only uses companies running on sustainable lines. Through its sister company ABAKOS, a similarly constructed service can be offered for other destinations. As she is against mass tourism, Director Suzi Stembridge individually plans every holiday.

Further information from: Sourdock Hill, Barkisland, Halifax, HX4 0AG. Tel. 0422 375999 Fax. 0422 310340.

ISLAND HOLIDAYS.

A special interest company which takes escorted tours to a number of worldwide destinations. Many of these tours are to destinations where the wildlife and natural environment play a crucial role in the traveller's needs (e.g. Falklands, Shetland). As such, Island Holidays recognises the need for environmental protection and all tours are led by naturalists or an archaeologist.

Further information from: Ardross, Comrie, Perthshire, PH6 2JU. Tel. 0764 70107 Fax. 0764 70958. (See p. 155 for detailed entry.)

MOSWIN TOURS.

This is a small company, established in 1981 by a German owner and specialises in holidays to Germany. Low impact and well thought of in the travel trade. For '92 includes Eastern Germany on homestays and stagecoach rides through Bavaria.

Further information from: 21 Church Street, Oadby, Leicester, LE2 5DB. Tel. 0533 714982 Fax. 0533 716016.

NSR TRAVEL.

The leading UK based travel agent/tour operator to Norway and Scandinavia. NSR Travel recognises that the natural environment plays a dominant role in clients' enjoyment of its holidays. As such NSR is committed to the principles of sustainable tourism and intends to provide its clients with an environmental checklist produced in conjunction with Green Flag International.

Further information from: Norway House, 21–24 Cockspur St, London, SW1Y 5DA. Tel. 071 930 6666 Fax. 071 321 0624.

PURE CRETE.

The company offers locally-owned village houses in the foothills of the White Mountains. Run as a partnership between Wildlife Travel and Pure Crete, money is donated towards conservation and the company concentrates on naturalist tours of walking and wildflower viewing. Promotes 'sympathetic tourism' in keeping with environmental aims.

Further information from: Acorn House, 74–94 Cherry Orchard Road, Croydon, CR0 6BA. Tel. 081 760 0879.

SAGA HOLIDAYS.

Specialises in travel for the over-sixties. Offers coach tours, long stays, long haul, treks, city breaks, cruises, railway journeys, skiing and special interest breaks. Currently offer eight Natural World tours and holidays where clients can experience developing countries. Environmental initiatives include working with conservation groups in host countries, and in conjunction with the British Trust for Conservation Volunteers, running a series of practical conservation projects from woodland management to repairing walls and footpaths.

Further information from: SAGA Group Ltd, The Saga Building, Middelburg Square, Folkestone, Kent, CT20 1AZ. Tel. 0303 857000.

SIMPLY TRAVEL (INCORPORATES SIMPLY CRETE, SIMPLY TURKEY, SIMPLY CORSICA AND SIMPLY SKI).

For 1992 a specialist tours programme is available, covering fauna and flora. An excellent operator taking sustainable principles seriously. Simply Crete specialises, exclusively, in Crete. The small staff have first-hand knowledge of the island.

Further information from: 8 Chiswick Terrace, Acton Lane, London, W4 5LY. Tel. 081 995 3883 Fax 081 995 5346.

SUN BLESSED HOLIDAYS.

Operating exclusively in Jersey, this company actively promotes environmental issues in all its holidays to this island.

Further information from: Victoria House, Princes Road, Ferndown, Bournemouth, Dorset, BH22 9JG. Tel. 0202 861616 Fax. 0202 861414.

SUNVIL HOLIDAYS.

Independent specialists operating to Greece, Cyprus, Italy and Portugal. Launched an extensive programme offering holidays in traditional village settlements restored by the National Tourist Organisation of Greece.

Further information from: 7–8 Upper Square, Old Isleworth, TW7 7BJ. Tel. 081 568 4499 Fax. 081 568 8330.

TRAVEL CLUB OF UPMINSTER.

A long-established family business, offering inclusive tours to the less developed resort areas in Majorca, the Algarve, Crete and the lakes and mountains of Italy, Austria and Switzerland. They offer both hotel and self-catering holidays and also a golf programme.

Paul Chandler, the company's Marketing Director and son of the firm's founder, states that since joining the Green Flag initiative he has instigated a green audit of the company's operations. A small but significant gesture in the first few months was to replace the plastic baggage labels with recycled cardboard ones.

Further information from: Station Road, Upminster, Essex, RM14 2TT. Tel. 04022 25000 Fax. 04022 29678.

TURKISH DELIGHT HOLIDAYS.

Independent specialist in holidays to Turkey with sound green travel ideas. Specialists, using low-impact facilities. Comprehensive coverage of country in many forms of travel.

Further information from: 164b Heath Road, Twickenham, TW1 4BN. Tel. 081 891 5901.

VFB HOLIDAYS.

Established in 1970, VFB Holidays is a fully bonded, independent tour operator specialising in opening up rural France to the British visitor. Still best known as the pioneers of gîte holidays, VFB now offers a wide range of programmes including Alpine activity, canal, short break and hotel-based holidays in mainland France, Corsica and the French Caribbean. The company has a policy of working with only locally resident French nationals. Holidays can be arranged in the Corsican village of Lama, a small mountain community which suffered considerable depopulation during the agricultural decline of the sixties and seventies, but which has now been revitalised by its enthusiastic inhabitants. Visitors are encouraged to share village life and invited to assist in community projects involving, perhaps, the preparation of a *son et lumière* performance or folk festival.

Further information from: Normandy House, High Street, Cheltenham, Gloucestershire, GL50 3HW. Tel. 0242 580187 Fax. 0242 570340.

WILDLIFE TRAVEL.

Wildlife Travel was set up in 1987 to raise funds for RSNC The Wildlife Trusts Partnership. It aims to organise travel programmes for naturalists from overseas to visit nature reserves and areas of natural beauty in the UK, and to take parties from this country to learn more about the culture and wildlife of other countries, especially in the Mediterranean. Staff and guides all have a deep concern for wildlife conservation and the company is currently engaged in protecting sites in Crete and Cyprus by encouraging sympathetic tourism by people who care about the long-term future of the areas they are privileged to visit. Accommodation is normally in 3- and 4-star hotels with daily excursions by small coach or on foot. Prices vary from £450 for a week in Crete to £700 for ten days in Cyprus.

Further information from RSNC, Vigilant House, 120 Wilton Road, London, SW1V 1JZ. Tel. 071 931 0601 Fax. 071 828 6297.

ZOE HOLIDAYS.

Zoe offer alternative tours to rural Greece. 'The tours will appeal to those seeking to see something of contemporary life in Greece and wishing to obtain a wider historical perspective than that offered by tours concentrating on ancient sites.' Visits places off the main tourist map. Small groups (under 18); extensive use of locals as guides. Interesting diverse itineraries.

Further information from: 34 Thornhill Road, Surbiton, KT6 7TL. Tel. 081 390 7623.

NATIONAL TOURIST BOARDS

We feel it pointless to rate *countries* in terms of 'green-ness', for there can be no properly comprehensive criteria applied to what a national tourist board tells you they're doing and what they *are* actually doing (and it's impossible to go out and verify a whole country's environmental tourist policy!), but

Conventional Holidays 235

suffice to say that those making the right noises, whether by issuing credible-looking policy documents, or by joining Green Flag International, are: Canada, Denmark, Falklands, Malta, Mexico, Norway, St Vincent & the Grenadines and Switzerland. The Caribbean countries are particularly interested in developing Eco-tourism.

Chapter Notes

Chapter 1
1. World Tourism Organisation, *Compendium of Tourism Statistics*, tenth edition, WTO, 1991.
2. Department of Employment, *Tourism and Travel Statistics*, HMSO, August 1990.
3. Hunziker, W., *Revue de Tourisme*, Berne, Switzerland, 1961.
4. Katie Wood interview with Martin Brackenbury, Deputy Director, Thomson Holiday Group, 18 September 1990.

Chapter 2
1. WTO, *Compendium of Tourism Statistics*, tenth edition, WTO, 1989.
2. United Nations Environment Programme, *Mediterranean Action Plan*, Mediterranean Co-ordinating Unit, 1987.
3. 4, 5. See note 1.
6. Lea, J., *Tourism and Development in the Third World*, Routledge, 1988.
7. O'Grady, A. (Ed.), *The Challenge of Tourism*, The Ecumenical Coalition on Third World Tourism, 1990.
8. Department of Employment, *Tourism and Travel Statistics*, HMSO, August 1990.
9. Wyer et al, *The UK and Third World Tourism*, TEN publications, 1988.

Chapter 3
1. Department of Employment, *Tourism and Travel Statistics*, HMSO, August 1990.
2. Wyer et al, *The UK and Third World Tourism*, TEN Publications, 1988.
3. Lea, J., *Tourism and Development in the Third World*, Routledge, 1988.
4. O'Grady, A., *The Challenge of Tourism*, ECTWT, 1990.
5. Wyer et al, *op. cit.*
6. *New Internationalist* magazine, December 1984.
7. *UNEP State of the Mediterranean Marine Environment*, MAP Technical Report Series No. 28, UNEP, Athens, 1989.

8. Mojetta, A., 'Recipe for a Slimy Sea', *World* magazine, June 1990.
9. Mason, P., *Tourism: Environment & Development Perspectives*, World Wide Fund for Nature, 1990.
10. UNEP, 'Carrying Capacity for Tourism Activities', *UNEP Industry and Environment*, Vol. 9, no. 1, 1986.

Chapter 4
1. United Nations Environment Programme, *Tourism*, UNEP, 1982.
2. Rice, Alison, 'Crisis on the Crumbling Costas', *Observer*, 4 February 1990.
3. Claver, Arturo, 'The Greening of Spain', *Observer*, 11 February 1990.
4. O'Grady, A., *The Challenge of Tourism*, ECTWT, 1990.
5. UNEP, *op. cit.*, p.551.
6. Mason, Peter, 'Tourism – Environmental & Development Perspectives', World Wide Fund for Nature, 1990.
7. 'Demystifying Gorillas', *Green* magazine, Summer 1990.
8. McNeedy, J.A., & Thorsell, J.W., 'Jungles, Mountains & Islands: How Tourism can help Conserve the Natural Heritage', *World Leisure & Recreation*. Vol. 31, note.
9. Keating, B., 'The Secret Life of Tigers', *Travel & Leisure*, Sept. 1990, p. 146.
10. 'Tourism the Destroyer?', *Holiday Which?*, March 1990.

Chapter 5
1. Elkington, J., Burke T., Hailes, J., *Green Pages* 1988.
2. Attenborough, D. *State of the Ark*. Collins/BBC Books, 1987.
3. John Hooper, 'The Highway to the Spanish Byways', *Guardian*, 10 August 1990.
4. O'Grady, A. *The Challenge of Tourism*, ECTWT, 1990.
5. Pearce, D., Markandya, A., Barbier E.B., *Blueprint for a Green Economy* Earthscan Publications, 1990.
6. Krippendorf, J., *The Holidaymakers*, Heinemann, 1987.
7. EEC (1990), Tourism and Regional Development: Economic and Social Assembly, Brussels.
8. English Country Cottages Survey, 1990.
9. O'Grady, A. op cit.
10. English Tourist Board (1991), Tourism and the Environment: Maintaining the Balance, ETB/Employment Group.
11. Tyler C.,' Saving Singapore's Soul', *World* Magazine, Oct. 1990.
12. *Observer* 10 Feb. 1991, Jenny Woolf, 'The Greening of Mickey Mouse.'

Chapter 6
1. O'Grady, A. *The Challenge of Tourism* ECTWT, 1990.
2. British Tourist Authority Statistics, 1989.
3. *Tourism Concern* Newsletter No.3, p.11.

Further reading for the good tourist

Attenborough, D. *The First Eden: the Mediterranean World and Man*, Collins/BBC Books, 1991.
Barrett, F. *The Independent Guide to Real Holidays Abroad*, Newspaper Publishing Group, 1991.
Button, J. *Green Pages*, Optima, 1990.
Caines, R. (ed) *The Good Beach Guide*, Marine Conservation Society, Ebury Press, 1991.
Central Bureau for Educational Visits and Exchanges, *Volunteer Work*, 1989.
Central Bureau for Educational Visits and Exchanges, *Working Holidays*, 1991.
Davidson, R. *Tourism*, Pitman Publishing, 1989.
Edington, J.M. & M.A. *Ecology, Recreation & Tourism*, Cambridge University Press, 1988.
English Tourist Board, *Tourism and the Environment: Maintaining the Balance*, Glasgow & Associates, 1991.
Frommer, A. *New World of Travel*, Prentice Hall, 1990.
Grunnfield, F. *Wild Spain*, Ebury Press, 1988.
Hatt, J. *The Tropical Traveller*, Pan Books, 1985.
Hawkins, D.E., & Brent Richie, J.R. (eds), *World Travel and Tourism Review: Indicators, Trends and Forecasts*, Vol. 1, C.A.B. International, 1991.
Krippendorf, J. *The Holiday Makers*, William Heinemann, 1989.
Lea, J. *Tourism and Development in the Third World*, Routledge, 1988.
Mason, P. *Tourism: Environment and Development Perspectives*, World Wide Fund for Nature UK, 1990.
Millman, R. *The Responsive Traveller's Handbook*, CART, 1990.
National Tourist Boards of England, Scotland, Wales and Northern Ireland, *Visit Britain at Work*, Visitor's Publications, 1991.
O'Grady, A. (ed) *The Challenge of Tourism*, The Ecumenical Coalition on Third World Tourism, 1990.
Pearce, D., Markandya, A., Barbier, E.B. *Blueprint for a Green Economy*, Earthscan Publications, 1989.
Turner, L., Ash, J. *The Golden Hordes: International Tourism and the Pleasure Periphery*, Constable, 1975.
UK Tourism and Holiday Travel Market Review, Keynote Publications, 1989.

United Nations, *Mediterranean Action Plan*, Mediterranean Co-ordinating Unit U.N.E.P., 1985.

United Nations, United Nations Environment Programme 'Tourism – World Trends', pp. 545–559, 1982.

Wyer, J., Turner, J., Millman, R., Hutchison, A. *The UK and Third World Tourism*, TEN Publications, 1988.

Appendix

Contact Organisations

ARK TRUST.
Ark aims to make individuals aware of their personal impact on the natural world and offers information on environmentally safe products. It promotes campaigns to encourage recycling and positive action by local communities. Innovations include a range of environment-friendly household products available from most major supermarkets. A recent initiative has been Ark's 'Green Travel Bug Campaign' targeted at holidaymakers to help them take simple, sensible precautions to enjoy themselves whilst still protecting the environment and culture of their holiday destination. The campaign, supported by and launched at Manchester Airport, consists of in-flight video on aircraft and departure lounges, a children's environmental activity pack, magazine and 'Travel Bug' merchandise.

Further information from: 498–500 Harrow Road, London, W9 3QA. Tel. 081 968 6780 Fax. 081 968 6355.

ASSOCIATION OF BRITISH TRAVEL AGENTS (ABTA).
Formed in 1950, ABTA is the organisation which represents 90 per cent of tour operators and travel agents in Britain. ABTA's principal duties are to create as favourable a business environment as possible for its members while ensuring that standards of service and business practice throughout its membership meet the laid-down standards and codes. It is a self-regulatory body which is run by its membership.

ABTA has recently faced a number of problems related to the impending introduction of the EEC regulations to the travel industry plus the failure of members such as ILG. It has not formed any strong views on the impact of tourism on the environment though this approach is a possible method of improving standards within the industry.

Further information from: 55–57 Newman St, London, W1P 4AH. Tel. 071 745 7261 Fax. 071 637 0713.

BELLERIVE FOUNDATION.

An organisation founded by Prince Sadruddin Aga Khan which recently launched – in February 1990 – the Alp Action campaign to tackle pollution, erosion and deforestation in the Alps. Initiatives include a 'code of ethics' credit-card, summarising codes of conduct for tourists in the Alps, and a prototype nature reserve to integrate tourism and conservation.

Further information from: Alp Action, PO Box 6, CH-1211, Geneva 3, Switzerland. Tel. 41-22-468866 Fax. 41-22-479159.

BRITISH ACTIVITY HOLIDAY ASSOCIATION (BAHA).

Formed in 1986, the BAHA deals with virtually anything of relevance to activity holidays in Britain. Top of its list of 'Objects and Powers' is to 'maintain standards of safety, instruction and quality of activity and special interest holidays', followed by 'to provide a network for the exchange of ideas and information on activity and special interest holidays.' Other projects which the association may be willing to undertake include the organisation of conferences and meetings, the dissemination of information and the promotion and execution of, or assistance in, research surveys and investigations. Membership is open to anyone who provides 'activity or special interest holidays of whatever nature and has done so for a minimum of twenty-four months.'

Further information from: Rock Park, Llandrindod Wells, Powys, Wales, LD1 6AE. Tel. 0597 3902.

BRITISH ASSOCIATION OF TOURIST OFFICERS (BATO).

BATO is an organisation of public sector tourism officers who meet and discuss mutual problems. From time to time these may concern environmental issues, however the association makes no pretence at being an environmental organisation, being simply a mouthpiece for its members.

Further information from: c/o Plymouth Marketing Bureau, St Andrews Court, 12 St Andrews Street, Plymouth, Devon, PL1 2AH. Tel. 0725 261125.

BRITISH RESORTS ASSOCIATION (BRA).

The BRA is the only association open to all local authorities (regardless of type of authority), Tourist Boards and similar organisations within the UK, Isle of Man and Channel Islands who have a strong commitment to the promotion and development of tourism. The association is funded entirely from members' subscriptions and presents an entirely independent voice for tourism within the UK. Issues debated recently cover everything from resort entertainment to the problems of graffiti and vandalism; the holiday complaints procedures and the formation of the single European market in 1992. Members are located throughout the country, from Cornwall to Aberdeen and meetings are held three times a year at chosen member venues.

Further information from: PO Box 9, Margate, Thanet, Kent, CT9 1XZ. Tel. 0843 225511 Fax. 0843 290906.

BRITISH TOURIST AUTHORITY (BTA).
Promoting tourism to and within Britain, the BTA has representatives throughout the world. The association is also responsible for advising the government on tourism matters in general, as well as ensuring adequate provision of facilities and amenities of an acceptable standard throughout Britain.

Further information from: Thames Tower, Black's Road, Hammersmith, London, W6 9EL. Tel. 081 846 9000 Fax. 081 563 0302.

CENTRAL COUNCIL FOR EDUCATIONAL VISITS AND EXCHANGES.
Dealing primarily with the under-thirty-fives and working from the same address as the Youth Exchange Centre, the Central Council for Educational Visits and Exchanges is developing a widening range of opportunities for responsive travel in many shapes and guises. The council also publishes various useful guides such as *Working Holidays Abroad* and *How to Make the Best Use of Teacher Exchange* and details the work of organisations such as Project Europe (travel bursaries for young people in Europe); school-linking and pen-friend services and both the Arts and Sports Councils.

Further information from: Seymour Mews House, Seymour Mews, London, W1H 9PE. Tel. 071 486 5101.

CENTRE FOR THE ADVANCEMENT OF RESPONSIVE TRAVEL (CART).
The centre, run as a personal consultancy by Dr Millman, acts as a clearing house for information on 'responsive travel' and produces a leaflet, 'Guide Notes for Responsive Travel – Credo for the Caring Traveller', as well as a more substantial publication, *The Responsive Traveller's Handbook – a Guide to Ethical Tourism Worldwide*' (price £5).

Further information from: Dr Roger Millman, 70 Dry Hill Park Road, Tonbridge, Kent, TN10 3BX. Tel. 0732 352757.

CONVENTION OF INTERNATIONAL TRADE IN ENDANGERED SPECIES OF WILD FLORA AND FAUNA (CITES).
CITES came into force in July 1975 and now has over ninety member countries. The convention is dedicated to putting an end to over-exploitation of the wildlife trade and to prevent international trade from threatening species with extinction. Working closely with the World Wide Fund for Nature, these aims form the major components of the World Conservation Strategy launched in 1980 by the United

Nations Environment Programme (UNEP), the International Union for Conservation of Nature and Natural Resources (IUCN) and the World Wide Fund for Nature. CITES is funded by contributions from member states but is always seeking additional funding for a wide range of special projects, such as the preparation of an identification manual to assist customs officers in recognising specimens of protected species.

Further information from: Conservation Monitoring Centre (CMC): 219c Huntingdon Road, Cambridge, CB3 0DL. Tel. 0223 277314.

COUNTRYSIDE COMMISSION.
Preserving England's natural beauty is the aim of the Countryside Commission. This is achieved by advising the government on countryside policy, designating national parks, advising on landscape conservation and promoting the Access Charter and country code to name just a few activities. The Countryside Commission often works in conjunction with public bodies and has grants available for projects dedicated to conserving and improving countryside recreation and access facilities.

Further information from: John Dower House, Crescent Place, Cheltenham, Gloucestershire, GL50 3RA. Tel. 0242 521381.

COUNTRYSIDE COMMISSION FOR SCOTLAND.
The Scottish counterpart of the Countryside Commission, the organisation plays a similar role as its English counterpart, aiming to conserve and improve the Scottish landscape wherever possible. The commission advises planning authorities and the Secretary of State for Scotland and gives financial assistance for tree planting, countryside recreation and access facilities. It also subsidises the Countryside Ranger Services and Ranger Training Courses.

Further information from: Battleby, Redgorton, Perth, Scotland, PH1 3EW. Tel. 0738 27921.

DEPARTMENT OF EMPLOYMENT.
The Department of Employment has responsibility for domestic tourism policy in England and for the UK as a whole, though local policy for Scotland, Wales and Northern Ireland lies with the respective Secretaries of State. There is a Minister for Tourism, currently Viscount Ullswater, at Parliamentary Under-Secretary of State level. The department collects statistics and disseminates information on tourism, shapes government policy and oversees the activities of the tourist authorities. The Department recently launched a 'green tourism initiative' to identify the best practice for tourism development in harmony with the environment and ultimately to draw up guidelines for the industry as a whole.

Further information from: Caxton House, Tothill Street, London, SW1H 9NF. Tel. 071 273 5806/7 Fax. 071 273 5821.

DEPARTMENT OF THE ENVIRONMENT, ENDANGERED SPECIES BRANCH.
In conjunction with HM Customs and Excise, the branch is responsible for overseeing the trade in wildlife, plants and associated goods made from endangered species. It produces two leaflets, 'Endangered Plants' and 'Endangered Species'.

Further information from: Tollgate House, Houlton Street, Bristol, BS2 9DJ. Tel. 0272 218 202.

EARTHWATCH.
Earthwatch works throughout the world matching enthusiastic paying volunteers with scientists who need funds and manpower (see also Chapter 7).

Further information from: Belsyre Court, 57 Woodstock Road, Oxford, OX2 6HU. Tel. 0865 311600.

ECUMENICAL COALITION AGAINST CHILD EXPLOITATION.
ECCE is the national group for Britain and Ireland dedicated to tackling the problem of child exploitation in sex tourism. The aim is to raise the issues in those countries which send tourists to places like Thailand and the Philippines and campaign for the ending of such sex tourism.

Further information from: 57 Sharman Road, Northampton NN5 5JZ.

ECUMENICAL COALITION ON THIRD WORLD TOURISM (ECTWT).
Formed in 1982 and based on earlier work done by the World Council of Churches on Tourism, the coalition consists of a number of churches and associations who share a common concern about tourist development in the Third World. They aim to encourage awareness about issues and exploitative practices in the tourist industry and to be able to work together in creating tourist development on an equal basis for all those involved. The ECTWT welcomes personal visitors to its Bangkok HQ and produces *Contours* magazine, available by subscription.

Further information from: PO Box 24, Chorakhebua, Bangkok 10230, Thailand. Tel. 010 662 510 7287.

ELEFRIENDS.
Campaigning against the culling of elephants and the ivory trade.

Further information from: 162 Boundaries Road, London, SW12 8HG. Tel. 081 682 1818.

ENGLISH TOURIST BOARD.
The English Tourist Board provides incoming visitors with a comprehensive introduction to the country. Information on hotels, tours and travel companies throughout England is available. The Tourist Board

is also concerned with promoting England as a tourist destination and with maintaining the standards of facilities and amenities.

Further information from: Thames Tower, Black's Road, Hammersmith, London, W6 9EL. Tel. 081 846 9000 Fax. 081 563 0302.

ENVIRONMENTAL TRANSPORT ASSOCIATION.

Launched in April 1990 and receiving support from the WWF service, a campaigning organisation representing transport users who are concerned about the environment. The ETA wants to see priority given to the most environmentally friendly forms of transport such as walking, cycling, trains and buses, and protection of our natural environment and national heritage from the damaging effects of transport. ETA provides all the usual benefits of a motoring organisation, such as road rescue, helpline, insurance cover, car rental discounts, together with producing a bi-monthly magazine, *Going Green*, which examines transport and environmental matters.

Further information from: 17 George St, Croydon, CR0 1LA, Tel. 081 666 0445 Fax. 081 666 0422.

EUROPA NOSTRA.

Founded in 1963, Europa Nostra is a 'federation of more than 200 independent conservation organisations and 100 local authorities in twenty three European countries.' The organisation's aims are, 'To protect Europe's national and cultural heritage; to encourage high standards of town and country planning and architecture; to improve the European environment and to increase awareness of Europa Nostra in order to generate membership among a wider public.' Over the past few years Europa Nostra has influenced the Greek government in their decision to build an alumina plant away from Delphi (1985); supported the Council of Europe in their opposition to proposals to build a major road in the vicinity of Pompeii (1987), and helped in passing a resolution opposing the proposal to have the Region of Veneto as the location for the World Exhibition in the year 2000, in view of the pressure it would exert on the city and its citizens (1990).

Further information from: Lange Voorhut 35, 2514 EC, The Hague, Netherlands. Tel. 010 317 035 17865.

EUROPEAN BLUE FLAG.

This organisation runs the Blue Flag awards for beaches which have grown in prestige and public awareness in recent years and have served to highlight the unhealthy state of many of Britain's beaches. In order to be awarded a Blue Flag beaches must have a high standard of water quality; be cleaned daily during the busy season and have good facilities such as toilets, first-aid and life saving equipment.

Further information from: c/o Tidy Britain Group, 10 Barley Mow Passage, Chiswick, London, W4 4PH. Tel. 081 994 6477.

EUROPE CONSERVATION.
This is a non-profit making environmental organisation aiming to spread knowledge about Europe's natural heritage. It does this by organising holidays and research programmes in parks and nature reserves throughout Europe. Members can participate as volunteers in ecological and archaeological research.

Further information from: via Fusetti, 14, 20143 Milano, Italy. Tel. 39 2 5810 3135 Fax. 39 2 8940 0649.

FIELD STUDIES COUNCIL.
In its 'Environmental Ethic' the FSC says it is 'committed to promoting an environmental ethic in its own operations where possible. This involves policies of energy conservation in its buildings, sensitivity to the impact of fieldwork on field sites, conservation and ecologically sensitive land management on its estates and grounds, the use of environmentally friendly products in its units and the efficient use of energy and natural resources in catering.' The FSC concerns itself largely with education, offering courses at centres throughout the country, many of which are housed in properties leased from the National Trust. All centres are well equipped and situated in areas which are ideal for field study weekends. In addition courses/expeditions are offered overseas, from Canada to Swaziland. Subjects cover a wide range, from study of birds and flowers to painting and photography.

Further information from: Central Services, Preston Montford Hall, Montford Bridge, Shrewsbury, SY4 1DX. Tel. 0743 850380.

FRIENDS OF THE IONIAN.
A group formed under the leadership of the Greek Isles Club who operate holidays in these islands. Their aim is to preserve the natural beauty of the environment and they adopt practical measures such as beach cleaning and help to conserve the turtles of Zakynthos and the Mediterranean monk seals.

For further information: 66 High Street, Walton on Thames, Surrey, KT12 1BU. Tel. 0932 247617.

GENERAL DIRECTORATE OF THE ENVIRONMENT.
Controls the European Environment Agency and shapes EEC policy on environmental matters, including problems related to tourism development and natural heritage.

Further information from: Directorate General XI, Commission of the European Community, 200 rue de la Loi, 1049 Brussels, Belgium. Tel. 00-32-2-235-11-11 Fax. 00-32-2-235-01-44.

Appendix 247

GREEN FLAG INTERNATIONAL.

Green Flag International is a private non-profit making company limited by guarantee, formed 'in response to the growing demand for conservation advice from tour operators and the travelling public.' Its primary aim is to 'work in partnership with the tourism industry to make improvements to the environment worldwide.' Green Flag is financed through an initial one-off fee of £150 per operator, as well as through a public search fee of £5 and contracts for environmental advice from individual operators. All profits are distributed through a charitable trust to international conservation projects and organisations at tourist destinations. Membership is open to 'any company which is seeking to improve the environmental quality of its operation or holidays.' (See chapters 5 and 12.)

Further information from: PO Box 396, Linton, Cambs, CB1 6UL.

GREENPEACE.

Although Greenpeace are not renowned for their involvement with tourism, many of their campaigns touch upon issues relevant to the subject, notably their work with the loggerhead turtles in Greece. Greenpeace describe themselves as, 'An international independent environmental pressure group which acts against abuse to the natural world.' They say they tackle the threat to wildlife from direct killing, pollution and habitat loss and that the organisation 'faces the major dangers caused by the production and release into the environment of radioactive material and the threat of nuclear war.' Greenpeace has been at the forefront of environmental issues for almost twenty years now and say that their activities are backed by scientific research undertaken by skilled and qualified personnel. In Britain Greenpeace has over 270,000 supporters and roughly 3.5 million worldwide. There are Greenpeace offices throughout the world and they have a research base 'on the frozen continent of Antarctica'.

Further information from: 30/31 Islington Green, London, N1 8XE. Tel. 071 354 5100.

INTERCHANGE.

Interchange promote and organise professional, educational and cultural visits and exchanges throughout the world, catering for all ages. Their three basic programmes are 'Programmes in the United Kingdom', 'Programme Visits Worldwide' and 'International Exchange Projects'.

Further information from: Secretary, Lloyds Bank Chambers, 186 Streatham High Road, London, SW16 1BG. Tel. 081 677 9598.

INTERFACE – NORTH-SOUTH TRAVEL.

North-South Travel, working as part of Interface Tourism Association is one of the original travel agencies to combine 'cheap travel with commitment of its profits to development projects in so-called "Third World Countries", through its charitable trust.' North-South Travel has been established for over ten years and links with InterFace 'on issues of global tourism and opportunities for alternative travel'. The organisation will usually be able to get cheaper fares than most high street travel agencies, but states quite realistically that it does not claim to offer the guaranteed cheapest-possible tickets to all destinations, because the discounting competition is so fierce and search takes time and is therefore money-consuming. It does promise to do its best to find the most reasonable service for clients that can be obtained and which is consistent with the policy to pass on any profit to those in need, 'after meeting its own modest costs'.

Further information from: Moulsham Mill Centre, Parkway, Chelmsford, Essex, England, CM2 7PX. Tel. 0245 492882/872 Fax. 0245 356612.

Also working as part of the InterFace organisation is Travel Friends International. This is an organisation which aims to provide two-way relationships between people who travel, independently or in groups and people in the receiving countries. It does this by arranging 'before you go' weekends for intending tourists and travellers and people to people visits and excursions in-country. The first area in which this has been achieved is Goa where, as Canon Edward Finch, founder of the InterFace Ecumenical Academy points out, there has been 'confrontation between "charter tourism" and the offended people of Goa.' Contact is through Interface.

THE INSTITUTE OF TRAVEL & TOURISM.
The Institute is a professional body aiming 'to develop the professionalism of its members within the industry'. Membership is open to those appropriately qualified members of the travel and tourism industry. The Institute runs seminars and activities throughout the UK encouraging exchange of views and contacts for members.

Further information from: 113 Victoria Street, St Albans, Herts, AL1 3TJ Tel. 0727 54395. Fax 0727 47415

THE INTERNATIONAL AIR TRANSPORT ASSOCIATION (IATA).
IATA is the world trade organisation of scheduled airlines. Its members carry the bulk of the world's scheduled international air traffic under the flags of over a hundred nations. IATA's main purpose is to ensure that airline traffic worldwide moves with the greatest possible speed, safety, convenience and efficiency. Since they arise from the very basic necessities of international air transport, IATA's aims and activities are essentially practical. On its technical side IATA has an interest in developing airline policy on important environmental

Appendix 249

issues, including aircraft noise, aviation fuel standards and emissions from aircraft engines.

Further information from: 26 Chemin de Joinville, 1216 Cointriu-Geneve, Switzerland. Tel. 010 41 22 983366.

INTERNATIONAL FUND FOR ANIMAL WELFARE (IFAW).

Founded in 1969, IFAW works for the protection of animals particularly through its 'Seal Watch' programme in Canada. This programme is designed 'to replace lost income to the fishermen of the Magdalen Islands due to the EEC import ban of whitecoat products.' IFAW say that they are very concerned about the effect tourism may have on the seals and asked a group of independent Canadian scientists to study the interaction between seals and humans, which they did for two years. The conclusion is that there is virtually no negative effect at all and certainly no lasting effect. Other campaigns for the 1990s include persuading cosmetic companies to end animal experiments, encouraging South Korea to introduce tough anti-cruelty laws and supporting wardens in Uganda in their efforts to protect Africa's wildlife, especially the elephants. IFAW is funded entirely by public support and has over 650,000 supporters throughout the world.

Further information from: Tubwell House, New Road, Crowborough, East Sussex, TN6 2QH. Tel. 0892 663374.

MARINE CONSERVATION SOCIETY.

As well as campaigning for the protection of dolphins, seals, whales, shells, corals and virtually anything else connected with the marine world, the Marine Conservation Society is actively involved in creating and maintaining pollution-free beaches in the UK. In conjunction with the Coastal Anti-pollution League, they have launched a public awareness campaign to clean up Britain's beaches. Recent initiatives include a code for visitors to coral reefs ('Let Coral Reefs Live'), which is available free on application. The MCS also publish annually the *Good Beach Guide* which describes the best of British beaches and their current state of health.

Further information from: 9 Gloucester Road, Ross-on-Wye, Herefordshire, HR9 5BU. Tel. 0989 66017 and for the Coastal Anti-pollution League: Alverstoke, 94 Greenway Lane, Bath, BA2 4LN Tel. 0225 317094.

MEDITERRANEAN ACTION PLAN (MAP).

MAP came into existence following the 1972 United Nations Conference on the human environment. In 1975, during a conference called by UNEP, sixteen Mediterranean governments approved an Action Plan for the Protection of the Environment, which called for 'a series of legally-binding treaties to be drawn up and signed by

Mediterranean governments; the creation of a pollution monitoring and research network; and a socio-economic programme that would reconcile vital development priorities with a healthy Mediterranean environment.' Today MAP is carrying out extensive research on all aspects of the Mediterranean environment, concentrating on sustainable development policies which will ensure a healthier future for this area of the world.

The Blue Plan is part of MAP, analysing the various environmental issues, including pollution, tourism, energy and population, confronting Mediterranean countries. The Blue Plan was launched in 1979, as part of the Action Plan and explores the long-term evolution of the relationship between development and the environment. Its aim is to assist the Mediterranean states in making appropriate practical decisions for the protection of their marine and coastal environment. Public participation in this process is welcomed.

Further information from: 48 Vassileos Konstantinou Avenue, 116 35 Athens, Greece. Tel. 010 301 7244536/7236586.

NATURE CONSERVANCY COUNCIL (NCC).
As well as being a source of information on conservation issues, the NCC takes an active role in promoting conservation and raising public awareness. Members of the NCC have also worked in the travel industry as guides and advisors, notably with Cox and Kings.

Further information from: Northminster House, Peterborough, PE1 1UA. Tel. 0733 40345.

NORTHERN IRELAND TOURIST BOARD.
The Northern Ireland Tourist Board provides tourists with a comprehensive service, providing details of sights, accommodation, tours and travel. The board promotes Northern Ireland as a tourist destination, maintaining standards of facilities and amenities.

Further information from: 48 High Street, Belfast, Ireland, BT1 2DS. Tel. 0232 231221 Fax. 0232 240960.

ROYAL SOCIETY FOR NATURE CONSERVATION (RSNC).
The RSNC has been in existence for several years now. Its work is wide-ranging and plays an important role in raising environmental awareness. Recently RSNC has been involved in launching the British Wildlife Appeal; working with the Dorset Trust for Nature Conservation to raise money to buy 328 acres of meadows at Lower Kingcombe and at present is one of many organisations involved with UK 2000, working to make Britain's cities more 'environmentally friendly', restoring industrial heritage and developing new methods of controlling litter. The RSNC has also published a guide to Urban Wildlife Conservation entitled: *Green It Yourself: The DIY Handbook for Urban Wildlife Conservation*.

Further information from: The Green, Nettleham, Lincoln, England, LN2 2NR. Tel. 0522 752326.

ROYAL SOCIETY FOR THE PROTECTION OF BIRDS (RSPB).
The RSPB is the largest charity in Britain and one of the most influential conservation groups in the world. They have opposed development of Macedonia in Greece to save the silver pelican and, in the travel industry, are involved with many birdwatching companies who are members of the society, as well as organising holidays for individuals and groups.

Further information from: The Lodge, Sandy, Bedfordshire, SG19 2DL. Tel. 0767 80551.

SCOTTISH CONSERVATION PROJECTS.
Practical conservation in Scotland (see Chapter 7).
Freepost, Stirling, Scotland, FK8 2BR. Tel. 0786 79697.

SCOTTISH FIELD STUDIES ASSOCIATION.
Enochdhu, Blairgowrie, PH10 7PG. Tel. 025 081 286 (see Chapter 7).

SCOTTISH TOURIST BOARD.
The function of the Scottish Tourist Board is to provide visitors with details of tourist facilities within Scotland. It also aims to attract tourists to destinations in Scotland and to encourage the development of visitor facilities. In addition, the board maintains a discretionary system of grants and loans for tourism projects, including those for the improvement of visitor facilities and amenities in countryside areas.

Further information from: 23 Ravelston Terrace, Edinburgh, Scotland, EH4 3EU. Tel. 031 332 2433 Fax. 031 343 1513.

SURVIVAL INTERNATIONAL.
Campaigns for the basic human rights of tribal peoples, including fighting against inappropriate development and exploitation.

Further information from: International Secretariat, 310 Edgware Road, London, W2 1DY. Tel. 071 723 5535.

TOURISM CONCERN.
Conceived in 1988 as a British response and counterpart to various European groups, Tourism Concern acts as a network for linking people from all interests and backgrounds concerned about the nature of tourism development, particularly in the Third World. Membership costs £10, for which you receive a contacts list, free access to information, a newsletter and reduced prices on other available services such as video hire. Meetings and seminars are organised on a regular basis and look at particular aspects of the tourism business.

Further information from: Froebel College, Roehampton Lane, London, SW15 5PU Tel. 081 878 9053.

TOURISMUS MIT EINSICHT (TOURISM WITH INSIGHT).
A consortium of over twenty organisations (Tourism Concern is an affiliated member), Tourism with Insight is the foremost European forum for 'green' or 'gentle' tourism, with Professor Jost Krippendorf as its mentor.

Further information from: Arbeitsgemeinschaft Tourismus mit Einsicht, Herbert Hamele, Hadorfer Strasse 9, D-8130 Starnberg, Germany.

THE TRANSNATIONAL NETWORK FOR
APPROPRIATE (ALTERNATIVE) TECHNOLOGIES (TRANET).
TRANET was originally set up as a guide to people and organisations involved with developments of alternative technology throughout the world. Today, however, its quarterly newsletter has become an invaluable source of names and addresses of travellers interested in sustainable tourism, offering free accommodation and advice. Past issues have covered Australia and New Zealand, Japan and South-East Asia.

Further information from: PO Box 567, Rangeley, ME 04970, USA. Tel. 0101 207 864 2252.

THE TRAVEL AND TOURISM PROGRAMME.
The Travel and Tourism Programme is the first organisation of its kind in Britain, set up by some of the major service industries in conjunction with education authorities. Together they have developed a Travel and Tourism GCSE exam, piloted in London in 1986 and accredited in 1988. From September last year this was being run by around 180 schools and colleges throughout Britain. The partnership responsible for developing the Travel and Tourism Programme is now involved in associated curricular development both in the UK and abroad.

Further information from: The Director, 3 Redman Court, Bell Street, Princes Risborough, Aylesbury, Buckinghamshire, HP17 0AA. Tel. 08444 4208.

TRAIDCRAFT PLC.
'Traiding for a fairer world' is the motto under which Traidcraft promotes itself and works. Revolving around the ideas of 'fair shares, concern for people and care for the environment', Traidcraft deals in household items, small furnishings, foods, recycled paper, clothes, jewellery, toys, art and materials, much of which is produced by hand. Traidcraft trades directly with communities in the developing world and Traidfair shops and associated outlets are scattered throughout Britain. Traidcraft has also been running holidays to India to help support its work, but at present there is some uncertainty about whether these will continue.

Further information from: Kingsway, Gateshead, Tyne and Wear, NE11 0NE. Tel. 091 495 0591.

TRAVEL FRIENDS INTERNATIONAL. see INTERFACE – NORTH-SOUTH TRAVEL.

UNITED NATIONS ENVIRONMENT PROJECT.
The UNEP monitors and assesses changes in the physical state of the earth's environment. It undertakes management work in the fields of environmental law and protection and in economic and social development. It has tackled specific issues related to the impact of tourism and has promoted action plans to assess and tackle environmental degradation in the Mediterranean and the Caribbean as well as providing the secretariat for the CITES agreement.

Further information from: UK National Committee, 3 Endsleigh St, London, WC1H 0DD. Tel. 071 935 7160.

VOLUNTARY SERVICE OVERSEAS (VSO).
VSO have been working in some of the world's poorest areas for over thirty years. In that time they have placed more than 25,000 volunteers worldwide, usually responding to requests that include a strong training element. The organisation is a registered charity, 'dedicated to assisting less developed countries'. Volunteers can be from any walk of life, provided they have adequate training and a genuine interest in wishing to help. They must be prepared for a minimum two year commitment and VSO say 'experience and qualifications apart, the most essential qualities are sensitivity and adaptability'.

Further information from: 317 Putney Bridge Road, London, SW15 2PN. Tel. 081 780 2266 Fax. 081 780 1326.

WALES TOURIST OFFICE.
Promoting Wales as a tourist destination, the office also acts as a central information point for visitors for travel and accommodation within the country. In addition it maintains a general overview of standards of visitor facilities, encouraging improvement and development where necessary.

Further information from: Castle Street, Conwy, Gwynedd, Wales. Tel. 0492 592248.

WHALE AND DOLPHIN CONSERVATION SOCIETY (WDCS).
The WDCS is the only British charity which devotes all its resources towards the conservation and protecting of whales and dolphins. They say that, 'despite a major international campaign in the 1970s, the world's whales are still in desperate trouble. Recent surveys show previous estimates of whale numbers to be wildly optimistic. Incredibly, certain countries continue their whaling activities against the best

scientific and conservationist advice.' Membership with the WDCS costs £9 for a year (half price for the unemployed, and there is also a family membership of £12.50), for which you receive a twice-yearly newsletter/magazine, regular news updates and opportunities to join whale-watching expeditions.

Further information from: 20 West Lea Road, Bath, Avon, England, BA1 3RL. Tel. 0225 334 511.

WORLD CONSERVATION UNION.
Formerly known as the International Union for the Conservation of Nature and Natural Resources, the WCU aims to provide international leadership for the conservation and management of living resources. It provides advice and expertise on sustainable development as well as technical support to conservation treaties such as CITES. Several experts within WCU have contributed to the debate on the impact of tourism including ways in which National Parks can be compatibly managed for tourism and conservation, and the impact of man on the Alps.

Further information from: Avenue du Mont-Blanc, CH-1196 Gland, Switzerland. Tel. 010 4122 649114 Fax. 010 4122 642926.

WORLD TRAVEL AND TOURISM COUNCIL (WTTC).
The WTTC was only created in April 1990 by the Chief Executive Officers from a number of the world's leading travel and tourism companies. The council's stated role is to, 'research, package and deliver timely information and to work closely with existing industry associations to convince governments around the world of travel of tourism's importance to national economies.' Particular attention is to be paid to the creation of the single European market in 1992, as well as to the changes in Eastern Europe and other tourism issues around the world. The WTTC points out that it will give, 'support to protect the environment and strives to strike a responsible balance between respect for nature and tourism development.'

Further information from: Fifth floor, Windsor House, 50 Victoria Street, London, SW1H 0NH. Tel. 071 222 7966 Fax. 071 222 4983.

WORLD SOCIETY FOR THE PROTECTION OF ANIMALS.
The World Society for the Protection of Animals is a large anti-cruelty organisation, also involved with conservation and disaster issues. Some of their most visible work has been in the displays of anti-bull fighting posters seen in travel magazines and newspapers.

Further information from: 106 Jermyn Street, London, England, SW1Y 6EE. Tel. 071 839 3026.

Appendix

WORLD TOURISM ORGANISATION (WTO).

The WTO has its origins from 1925 at the First International Congress of Official Tourist Traffic Associations, which subsequently became the International Union of Official Travel Organisations and, in 1975, the WTO. Its aim is the, 'promotion and development of tourism with a view to contributing to economic development, international understanding, peace, prosperity and universal respect for and observance of human rights and fundamental freedoms for all, without distinction as to race, sex, language and religion.' The role of the WTO is to a large extent advisory, but it also conducts surveys and compiles statistics. It does not invest in tourist development directly.

Although the WTO has over 100 members, consisting largely of government tourism agencies and other affiliated tourism-related organisations, several countries, including the UK, are not members. This is due to a considerable dissatisfaction with what many see as the ineffectual role of the WTO.

Further information from: Capitan Haya 42, 28020 Madrid, Spain. Tel. 010 341 571 06 28 Fax. 010 341 571 37 33.

WORLD WIDE FUND FOR NATURE.

The world's largest nature protection organisation, WWF states that its aim is to, 'achieve the conservation of nature and ecological processes'. This can be achieved by using 'renewable natural resources', 'fighting waste of resources' and 'preserving ecosystem diversity'. Current tourism-related projects with which WWF is involved include 'Wetland Conservation: Menderes-Delta, Turkey' and 'Ecotourism in Northern Pindos'. As well as publishing information about itself, WWF also publishes several useful leaflets concerning animal and marine souvenirs including 'What on Earth Can I do?' advising how to minimise damage and pollution in everyday life. The WWF has recently highlighted its views on tourism development and conservation issues. As well as publishing a book on the subject, it has taken initiatives to promote sustainable tourism development in areas like Nepal. Another high-profile initiative has been the campaign to highlight the trade in souvenirs made from wildlife in general and endangered species in particular. Thailand, especially, has been targeted as one of the worst violators of the CITES convention, despite being a signatory.

Further information from: Panda House, Weyside Park, Catteshall Lane, Godalming, Surrey, GU7 1XR. Tel. 0483 426 444 Fax. 0483 426 409.

YOUTH HOSTELS ASSOCIATION (YHA).

The YHA is a registered charity, founded in 1930, concerned with educating the public about how best to conserve and care for the environment. Their hostels are spread throughout the world and as well as providing accommodation for anyone of limited means, are one of the original examples of 'community living' and tourist integration. The YHA offers a wide variety of holidays and special interest weekends, many of which focus on wildlife and natural history. The patron of the YHA is the Queen, and David Bellamy is its President.

Further information from: Department G, Trevelyan House, 8 St Stephen's Hill, St Albans, Herts, England, AL1 2DY. Tel. 0727 55215.

Also by Katie Wood and available in Mandarin

The Good Tourist in France

The number one destination for British holidaymakers, France is a country of extraordinary diversity. And as most people make their own way to France rather than taking a package, a good guidebook is an essential companion.

Following the success of *The Good Tourist* (which Jonathon Porritt called 'the definitive guide to the art of seeing the world without destroying it'), *The Good Tourist in France* is packed with completely new information and ideas to help you really enjoy France – without damaging it. By applying simple green principles you can radically reduce the impact that you, as a tourist, have on the places you visit.

We've asked the people who *really* know – the locals – for their inside information on holidaying in the UK:
- Where to go and where to avoid
- Ways to experience the local lifestyle
- Where to eat and stay in the best locally owned restaurants and hotels

Here is absolutely everything you need to be a 'good tourist', from Katie Wood, bestselling author and leading travel journalist, and Syd House, an experienced conservationist.

The Good Tourist in the UK

The UK – number one holiday destination for so many of us – is extraordinarily varied and exciting to explore. Whether you want to discover more about your own area or tour the whole country, this guide will show you how to get the very best from your trip.

Following the success of *The Good Tourist* (which Jonathon Porritt called 'the definitive guide to the art of seeing the world without destroying it'), *The Good Tourist in the UK* is packed with completely new information and ideas to help you really enjoy the UK – without damaging it. By applying simple green principles you can radically reduce the impact that you, as a tourist, have on the places you visit.

We've asked the people who *really* know – the locals – for their inside information on holidaying in the UK:
- Where to go and where to avoid
- Ways to experience the local lifestyle
- Where to eat and stay in the best locally owned restaurants and hotels

Here is absolutely everything you need to be a 'good tourist', from Katie Wood, bestselling author and leading travel journalist, and Syd House, an experienced conservationist.

The 1992 Business Travel Guide

In the world of business, 1992 either looms or beckons. Whatever happens, it will be a vitally important year.

The *1992 Business Travel Guide* contains all you need to know about a continent which is already changing fast – and how to enjoy it. Country by country, Katie Wood

- explains key business facts about politics, economics, tax laws, trading partners, customs and etiquette
- advises on special Business Class services, airport connections, high-speed rail links and local transport
- provides addresses of Embassies, Chambers of Commerce, reference libraries, 24-hour fax and photocopying facilities and essential foreign-language phrases
- and commends hotels, restaurants, nightclubs and venues for business entertaining as well as the best sights to see at all hours.

The 100 Greatest Holidays

Here for the first time is a compendium of the best holidays in the world. From the de luxe to the inexpensive, there is a chance for everyone to realise the holiday of their dreams:

- Great Journeys — The delights of travel for its own sake include the famous Orient Express, luxurious cruises and the 'Jules Verne' round-the-world tour

- Best Beaches — Sun worshippers can indulge in the sophisticated Seychelles, informal Phuket, the fantastic coral reefs on the Maldives or the atolls of the South Pacific

- Cities — Overtures to the great cities of the world – New York, Hong Kong, Sydney, San Francisco, Venice, Leningrad...

- Classic Tours — Choice tours of Tuscany, the Canadian Rockies, the Chateaux of the Loire or a jazz tour of the Deep South, to name but a few

- Adventure Holidays — Suggested itineraries for the independent in Mexico, Venezuela, Bhutan, Tibet; how to arrange African safaris or Arctic expeditions.

- Great Events — Experience the spectacular Carnival of Rio, the Songkran Festival in Bangkok or the Edinburgh International Festival.

- Activity Holidays — Trekking in Nepal, skiing in Colorado, exploring by hot air balloon or diving in the barrier reef – just a sample of the possibilities for the active

Whatever your interest, this book will guide you to the ultimate in holidays.